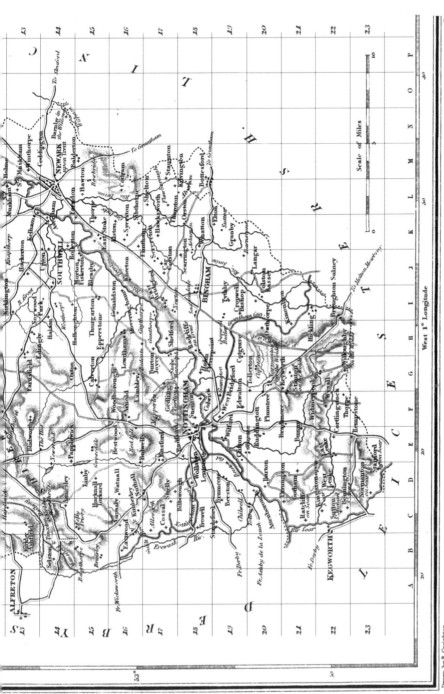

Scale of Miles

West 1° Longitude

Drawn by R. Creighton.

Unicorns

This book is dedicated to the 'Merry Men' of
Nottinghamshire and all Unicorns

*'The Unicorn is fierce yet good, selfless yet solitary, but always
mysteriously beautiful'*

Unicorns

The History of the Sherwood Rangers Yeomanry

1794–1899

Jonathan Hunt

Pen & Sword
MILITARY

First published in Great Britain in 2012 by
Pen & Sword Military
an imprint of
Pen & Sword Books Ltd
47 Church Street
Barnsley
South Yorkshire
S70 2AS

Copyright © Jonathan Hunt 2012

ISBN 978-1-84884-547-3

Typeset in 11pt Ehrhardt by
Mac Style, Beverley, E. Yorkshire

Printed and bound in the UK by CPI Group (UK) Ltd, Croydon, CRO 4YY

Pen & Sword Books Ltd incorporates the Imprints of Pen & Sword Aviation,
Pen & Sword Family History, Pen & Sword Maritime, Pen & Sword Military,
Pen & Sword Discovery, Wharncliffe Local History, Wharncliffe True Crime,
Wharncliffe Transport, Pen & Sword Select, Pen & Sword Military Classics,
Leo Cooper, The Praetorian Press, Remember When, Seaforth Publishing
and Frontline Publishing.

For a complete list of Pen & Sword titles please contact
PEN & SWORD BOOKS LIMITED
47 Church Street, Barnsley, South Yorkshire, S70 2AS, England
E-mail: enquiries@pen-and-sword.co.uk
Website: www.pen-and-sword.co.uk

Contents

Acknowledgements

I need to begin by confessing to the unconventional way in which this book has been both written and researched. The starting point, like a single bright but fluttering candle in a cavern, was a pamphlet compiled by Lieutenant Colonel D. C. Barbour, 17th/21st Lancers, and published in 1953 whilst he was commanding the Sherwood Rangers Yeomanry, covering the period from the raising of the regiment in 1794 to 1953 in 48 pages. It represented almost all the Regiment's collective knowledge of our history prior to 1900. In 1998 I set out to do no more than correct one or two minor points and update it to the present day.

I found myself looking for answers to the questions: 'Who, why and what was the context?' By the time I got the answers, albeit to only half the story, ten years had gone by and I am hard pressed to remember from whence my knowledge has come. In case you think ten years is rather a long time I should add in my defence that I found it to be a most rewarding, but time consuming hobby, and one which I only had time to indulge whilst on holiday.

I am certain there is more detail out there which I have missed, some of which could alter my interpretation of events or otherwise add to the story. I hope I have not got important facts wrong or have in some other way given offence to anyone for want of a key piece of information and apologize if I have. I would ask those who know more about aspects of the narrative than I do will forgive me and also correct me, and I will make sure the Regiment's archives are in turn corrected.

The best I can do at this stage is to try and produce a comprehensive list of my sources and the following, in no particular order is the result of my attempt to do so. Forgive me for any I miss:

The Archives of the Sherwood Rangers Yeomanry, which consist of correspondence and several comprehensive scrapbooks containing many newspaper cuttings from both the national and local press from the last half of the nineteenth century and the Regiment's collection of images; the two regimental histories of the South Nottinghamshire Hussars with whom the Regiment has shared so much over the years, not least parts of our history, and for permission to reproduce certain images. Many of the newspaper reports quoted in the early part of the narrative come from their research; the archives of the Earls of Scarbrough, especially applicable to the period covered by their ownership of Rufford Abbey and its estates, and in particular to Lord Scarbrough and his father for making available to me the numerous volumes of *The Complete Peerage* by G. E. C. Different in the library at Sandbeck, which I have trusted as accurate and also for certain images.

Thanks are also due to William Parente for his help; to Lord Middleton for his help; the archives of the Manvers family and for permission to reproduce certain images; the archives of the Galway family; the archives of the Woollaston White family and for several images; the archives of the Machin family including press cuttings and photographs.

A History of the English Speaking People by Winston S. Churchill helped to inform me of the 'bigger picture'. The internet also helped generally, particularly in cross checking information and adding detail; it is astonishing how much more is available now compared with when I started. For example, almost all the information about Thomas Wildman was sourced on the internet.

The University of Nottingham Library, much of it via the internet, was another valuable source. I also thank Bassetlaw Museum for scanning all the Sherwood Rangers archives and providing a number of images. David J. Knight and his valuable correspondence with me was also a great help.

Other valuable sources of information were Keith Douglas' book *Alamein to Zem Zem* and his poem 'Aristocrats' as well as the following books:

Benson Freeman's '*History of the South Nottinghamshire Hussars Yeomanry*'
The County Lieutenancy in the United Kingdom by C. Neville Packett
The Militia and Yeomanry Lists
Yeomanry Wars by Peter D. Athawes
Jean Loynes' Thesis on the Yeomanry in the nineteenth century
Where Truth Abides, the Diaries of the 4th Duke of Newcastle, edited by John
    Fletcher
Thomas Wildman's Letter Book
Young Tom Hall by R. S. Surtees
The Corps of Light Infantry 1758 in the French and Indian War
Socheage Hill to Sandbeck, edited by Philip Ireson
Sherwood Forest and the Dukeries by Adrian Gray

I would be remiss in not thanking Geoffrey Norton for reading and commenting on the pre proof and my son Edward Hunt's company, Hedgerow

Publishing Ltd for adapting the old map of Nottinghamshire for use in the book by lifting the old font used in the map itself to form missing place names and the grid. Thank you also to the team at Pen and Sword, in particular Henry Wilson whose sound advice caused me to rewrite this at least once and in parts twice, Richard Doherty my editor and Matt Jones on the selection of images and much else.

Last but not least thanks are due to my wife Sue who proofread the whole manuscript during a tour of New Zealand correcting the bits that made no sense, as she has been doing for forty years – not least, given I am a lawyer, adding punctuation, which helped a lot.

Illustrations

Front jacket: Sherwood Rangers Yeomanry Officer by Richard Simkin
Back jacket: Nottingham Castle on fire and 4th Duke of Newcastle-under-Lyne.

Page 30
Cover and page from *A Treatise on the new Broad Sword Exercise* by W. Pepper
  of the Nottinghamshire Yeomanry Cavalry.

Page 46 facing
Lieutenant Colonel E. O.Kellett and Lieutenant Colonel J. D. Player.
Lieutenant Colonel Commandant Anthony Hardolph Eyre.
Captain Sir Thomas Woollaston White,1st Baronet. (By kind permission of
  Sir Nicholas Woollaston White)
Cornet Alexander Hadden and the Nottinghamshire Troop during the Bread
  and Food Riots, 1795. (By kind permission of the South Notts. Hussars)
Captain the 2nd Earl Manvers. (By kind permission of the Manvers Trustees)
The late Jeremiah Brandreth, leader of the Derbyshire Rising, also known as
  the Brandreth Riots.
Captain Henry Charles Howard, Earl of Surrey and later 13th Duke of
  Norfolk KG. (By kind permission of the Duke of Norfolk)
Captain John Evelyn Denison, later 1st Viscount Ossington.
4th Duke of Newcastle-under-Lyne KG.
The Earl of Lincoln, later Lieutenant Colonel the 5th Duke of Newcastle-
  under-Lyne KG.
Lieutenant Colonel S. W.Welfitt.
Captain John Vessey Machin. (By kind permission of the Machin Family)
The 5th Duke of Newcastle entertaining HRH the Prince of Wales on the lake
  at Clumber Park (1861) Illustrated London News
HRH the Prince of Wales arriving at Clumber (1861) (Illustrated London News)
Clumber Troop escorting HRH the Prince of Wales through Worksop (1861).
  (Illustrated London News)

Worksop Manor. (By kind permission of Bassetlaw Museum)
Clumber Park. (By kind permission of Bassetlaw Museum)
Wollaton Park. (By kind permission of Lord Middleton)
Thoresby Park.

Page 110 facing

2nd Duke of Portland. (By kind permission of Mr William Parente)
3rd Duke of Newcastle-under-Lyne. (By kind permission of Nottingham
    University)
3rd Duke of Portland. (By kind permission of Mr William Parente)
Captain Ichabod Wright.
Trooper's uniform of the Nottinghamshire Yeomanry, 1798 by Charles
    Stadden.
1st Earl Manvers. (By kind permission of the Manvers Trustees)
Captain Richard Lumley-Savile,6th Earl of Scarbrough. (By kind permission
    of Lord Scarbrough)
5th Baron Middleton and family. (By kind permission of Lord Middleton)
Major the 4th Duke of Newcastle-under-Lyne KG.
6th Baron Middleton. (By kind permission of Lord Middleton)
Nottingham in the eighteenth century. (By kind permission of Lord Middleton)
Lieutenant Colonel Thomas Wildman.
Lieutenant Colonel, Sir Thomas Woollaston White, 2nd Baronet. (By kind
    permission of Sir Nicholas Woollaston White)
Nottingham Castle on Fire, 1831.
Lady Euphemia Woollaston White. (By kind permission of Sir Nicholas
    Woollaston White)
Vere, Viscountess Galway.
17th Yeomanry Brigade by Richard Simkin.
Officer and Trooper of the Sherwood Rangers Yeomanry 1840 by Charles
    Stadden.
Officer, Sherwood Rangers Yeomanry 1899 by Charles Stadden.
Lieutenant Colonel the 3rd Earl Manvers. (By kind permission of the
    Manvers Trustees)
The Regimental Standards.

Page 119

Nottingham Castle. (by kind permission of Charlotte White)

Page 149

Clumber Troop escorting HRH the Prince of Wales approaching Clumber
    (1861) (Illustrated London News)

Foreword

There is a photograph of the Sherwood Rangers Yeomanry at their summer camp in August 1939, lined up on their horses, boots polished and sabres at their sides. It is almost possible to hear the squeak and chink of leather and stirrups. This body of men, mostly country landowners drawn from the Nottinghamshire area, would soon be heading off to war, with their chargers. In the six years that followed, theirs would be an incredible odyssey – a journey through the Middle East, North Africa, through Northern France and across Northwest Europe and into Germany. And despite moments of ignominy, at the war's end, this extraordinary yeomanry regiment had a reputation for being amongst the finest cavalry units in the entire British Army, with, it is believed, more battle honours than any other.

The Sherwood Rangers Yeomanry was not only one of the most successful cavalry regiments of the war, it has also become one of the best known, largely because of the numerous memoirs written by former stalwarts of the regiment. Most famously of all, however, Keith Douglas, the celebrated war poet, who was killed with the regiment in 1944, not only wrote among the finest poems to emerge from the war, but also a classic memoir, *Alamein to Zem Zem*, about his time with the regiment in 1942–43.

Yet as Jonathan Hunt has so lucidly shown in this fine account of the first hundred or so years of the regiment's history, the Sherwood Rangers had played an important role within the British Army long before the Second World War plunged the world into global conflict.

Raised originally to help keep the peace in the Nottingham area at a time of civil unrest and revolution across the Channel, they continually proved themselves throughout the next hundred years, operating as yeomanry as part of an anti-invasion force against Napoleon and then helping maintain order during the various riots, reforms and civil protests that gripped Britain in the first half of the nineteenth century.

This in itself is fascinating, and in writing about the Sherwood Rangers, Jonathan Hunt is able to use them as a leitmotif not only for any yeomanry regiment at this time, but also to cast light on the rapidly changing social fabric of this period.

His research is impeccable, and extends far beyond the reaches of the Shire, for what makes this book so absorbing is the context that he brings to the narrative. There are many wonderful details, but also invaluable explanations, which makes this as much a general social history as an account of the birth and growth of the Sherwood Rangers Yeomanry. Weaving the micro with the macro is always a difficult balance to achieve, yet the story of the regiment with the wider, national and international picture is skilfully handled and the book is infinitely the richer as a result.

The work that Jonathan Hunt has done in this first volume of one of our finest yeomanry regiments is not only a wonderfully entertaining account, it is also an important work of historical research, and a testimony to all those who served in the tumultuous early years of its history.

James Holland

Preamble

Aristocrats

'I Think I Am Becoming A God'

The noble horse with courage in his eye,
clean in the bone, looks up at a shell burst:
away fly the images of the shires
but he puts the pipe back in his mouth.
Peter was unfortunately killed by an 88;
it took his leg away, he died in the ambulance.
I saw him crawling on the sand, he said
It's most unfair, they've shot my foot off.

How can I live among this gentle
obsolescent breed of heroes, and not weep?
Unicorns, almost,
for they are fading into two legends
in which their stupidity and chivalry
are celebrated. Each, fool and hero, will be an immortal.
These plains were their cricket pitch
and in the mountains the tremendous drop fences
brought down some of the runners. Here then
under the stones and earth they dispose themselves,
I think with their famous unconcern.
It is not gunfire I hear, but a hunting horn.

This poem was written by the war poet Captain Keith Douglas who was killed
in action with the Sherwood Rangers Yeomanry in Normandy, on 9 June 1944.

Keith Douglas wrote 'Aristocrats' in 1943 when he was twenty-three. He had
risen from a humble background on his own merits, but with help from those

who saw his talent, winning a place at Oxford where he had already been recognized as a writer and poet of great potential. Unusually he shared that interest with one for the military. Like many of his age and background he was bright, impatient of tradition, and unconventional.

On the outbreak of the Second World War he sought and gained a commission and was posted to North Africa. In one of those postings of opposites for which the fog of war, aided and abetted by Manning and Records, is renowned, he found himself in the Sherwood Rangers. They were in the throes of converting from horsed cavalry, via coastal gunners, to armour, following participation in the famous and successful first Siege of Tobruk and the unsuccessful defence of Crete. *Panzerarmee Afrika*, under Rommel, meanwhile, was pressing ever nearer to the Egyptian frontier and Cairo.

The reason why it was a posting of opposites was that the Sherwood Rangers were led by a group of officers from, to his mind, an aristocratic and very different background to his own; who seemed to Keith Douglas to exude an air of insouciance which left him impatient given the gravity of their situation. If to them he was merely irritating, they, of course, were disinclined to show it.

Before long he was to find, to his own irritation, that nothing changed in their demeanour when, now issued with their tanks, they became operational once more. Initially this was immediately before El Alamein, then prominently during the battle itself and thereafter. Save that both officers and men proceeded to fight their tanks in action after action on the 1,600 mile advance from Egypt to Tunis seemingly impervious to, and careless of, the extreme risks they were running. It was, of course, a façade.

An armoured regiment in action is controlled by means of a regimental radio net of upwards of sixty stations, including one for each tank with a second for the troop net. On a radio net only one message can be transmitted at a time; two or more simultaneous transmissions jam each other and nothing can be heard. If the net was to work strict discipline from everyone was essential since, at any given moment, at least one of those stations was likely to be either inflicting casualties or suffering them. In those circumstances air time became as precious as gold-dust, only the most urgent and important messages being permitted.

For the three hundred or so pairs of ears glued to it, the net magnified the intensity of the unfolding drama of battle and became the compelling personification of the regiment in action. The news it bore was capable of affecting morale; therefore, no matter what the content of a message, no one was expected to display emotion on the air. Everything had to sound matter of fact and normal, usually the opposite of the reality.

They also had to veil the meaning of their messages, often containing sensitive information, from an enemy who was able to monitor all they said. Radio security on the battlefield was in its infancy and they, incongruously, used obscure or humorous cricketing, hunting or racing metaphors and riddles as a primitive code. Keith Douglas found the combined effect of the matter-of-fact tone and the apparently irrelevant messages surreal in the crash of battle, particularly since he knew little of cricket and nothing of hunting or racing and was thus, in the early stages, in danger of being left as ill informed as the enemy.

They were one of three armoured regiments in a brigade whose role was, arguably the most dangerous of any formation in Eighth Army at that time. It was to locate and punch the initial hole through the enemy's defensive stop lines during the long advance. They did so at El Alamein and thereafter every time Eighth Army encountered resistance. Once they had done so, the following armoured divisions exploited the breach to harry the, by then, disrupted enemy. Inevitably many of the Regiment paid with their lives, for they were forever advancing against an enemy whose deadly 88mm anti-tank guns were hidden in ambush. In the process Keith Douglas learned at firsthand what it was like to have the armour of his tank penetrated by an armour-piercing shell from an 88. He likened it to 'the entry of a demon'. On that occasion he was merely wounded, but it is easy to see why it led him to the conclusion that anyone, himself included, who climbed back inside the hull of a tank on a daily basis, knowing the risks as they all did, must be both a fool and a hero. His admission to a fear that 'dawn might surprise me unpleasantly at my least heroic hour' expresses so graphically the state of mind of a soldier about to go once more into action.

The Regiment was checked, but never stopped. In the process they helped, more than most, to inflict attrition on the *Panzerarmee* and to drive it out of North Africa. In doing so they won a glittering reputation. They earned in return not only Keith Douglas's grudging respect and admiration, but his acknowledgement that he was succumbing to their ethos, madness even, which, as it transpired, would cost him his life.

Following the fall of Tunis, as a result of the high regard in which they were, by then held, the Sherwood Rangers were specially brought home. Their new task, their 'reward', was to form one of an elite group of four armoured regiments which were issued with 'amphibious' 34-ton tanks to spearhead, literally, the D Day landings on Gold and Sword, the two British beaches. Their role involved each tank being launched at sea to 'steam' shoreward in the vanguard, suspended beneath the waves under no more than a canvas screen supported by a few struts, and once on land tasked to provide close support to the infantry and assault engineers landing in the following waves. The Regiment's objective was the right-hand sector of Gold.

Two men above all impressed themselves on Keith Douglas. They were the two who commanded the regiment during the fighting in North Africa. Firstly, Colonel E. O. Kellett DSO MP (Flash), but also Lieutenant Colonel J. D. Player (Donny), a son of the tobacco family, who succeeded him.

Both, but mainly Flash Kellett, whose resolute qualities of leadership were the only exception to the 'economy of air time' rule, turned the Regiment into the battle-hardened unit it became. They paid, in turn, with their lives within a few weeks of each other and now, with many others 'lie under the stones and earth' of Tunisia. In the poem 'Peter' was Donny. Keith Douglas wrote 'Aristocrats', in honour and despair over the loss in action of yet a second commanding officer, as well as many close friends, shortly after Donny's death.

What he was too young to know was that both men and also their officers and yeomen had inherited their 'famous unconcern' from those from whom they had also inherited their regiment. They were the then current representation of an ethos of elitism and invincibility by right of birth that had been inculcated into the Sherwood Rangers' soul by surely one of the most extraordinary group of people ever to serve in or be associated with such a small, single battalion regiment.

When he wrote 'Aristocrats' he, unknowingly, wrote of them all, whilst they, for their part, in their own diverse ways, made history.

Chapter 1

Antecedents

The Sherwood Rangers Yeomanry originates from, and has always been based in Nottinghamshire, particularly, the area which embraces Sherwood Forest and which is also known as the Dukeries after four ducal families who historically had seats there.

It was formed as part of a cavalry regiment, called the Nottinghamshire Yeomanry, raised from the whole of Nottinghamshire, including Nottingham itself, in 1794 and which was reorganized into two units between 1826 and 1828, the Regiment being named the Sherwood Rangers Yeomanry becoming the senior of the two, covering most of the County, and the southern Troops being named the South Nottinghamshire Yeomanry Cavalry, and eventually, in 1887, being renamed the South Nottinghamshire Hussars Yeomanry and being based in and around Nottingham.

It is a volunteer reserve regiment and not a regular regiment.

The process by which the Regiment was raised and the nature of the organization that was created was part of an evolution which has been a keystone of the defence and survival of our nation for over 1,200 years, arguably, making that evolving process our oldest institution. The details had of course, varied over the centuries, and the process itself had from time to time, during periods of prolonged peace, fallen into abeyance. The thread was never lost, however, and the principles have remained constant:

- That the Monarch divided the country into areas, often but not always based on Shire boundaries.
- To each such area he appointed a loyal subject, often but not always a leading nobleman, to supervise the raising of volunteer forces when required for service in defence of the allotted area. In the early days this role was sometimes given to the High Sheriff, the country's oldest surviving secular appointment itself dating back a thousand years. The High Sheriff, the Shire reeve or scribe, originally had responsibility for almost everything in his Shire, from raising taxes, to law and order and general security, but even from

the earliest days there had been a preference to distinguish the role of raising such volunteer forces and treat it as a strictly military appointment. In due course this became universal and the title of the chosen individual evolved to that of Lord Lieutenant. In 1547 Francis Talbot, 5th Earl of Shrewsbury, the premier earldom of England, was in receipt from Henry VIII of the first permanent Lieutenancy Commission which was for the counties of York, Lancaster, Chester, Derby, Salop, and Nottingham. As mentioned, prior to that date individual noblemen had performed the role only in times of need. After that date, generally speaking, the majority of counties had someone permanently in appointment.

- Following the creation of a standing army at the time of the Restoration, the importance of the Lord Lieutenant was to some extent diminished in that he was no longer the only conduit through which armed forces were raised but even so he was still in full possession of the powers that he was originally given to raise the militia when the need to do so arose.
- The Lord Lieutenant remained fully empowered until 1871 when the Army Regulation Act was passed so that, for example, commissions into the militia and the yeomanry were signed by the Lord Lieutenant and not by the Monarch.
- To assist him in his task the Lord Lieutenant was entitled to appoint Deputies.
- The threats that these volunteer forces were intended to meet were limited to those of invasion, riot or insurrection. As originally conceived they were not intended, and neither were they used, for service overseas; their role was defensive. Normally there was no standing army at all and the defence of the country was entirely in the hands of whatever forces could be raised by the Lords Lieutenant or their predecessors.
- There was a need, from time to time, to raise troops for service overseas. This process evolved separately, and was of an intermittent and strictly temporary, need driven nature. However, when recruiting, it drew on the same pool of people, whom the Monarch reached through the landowners, although the purpose was different. This process did not result in any permanent structure until the Restoration of the Monarchy in the mid-seventeenth century. Thereafter a standing army has always existed to perform that role, which has evolved into the professional regular Army that we know and deeply respect today. Many of the regular regiments which exist today trace their origins to that time or even slightly earlier. The deployment of these regular troops, whether before or after the Restoration, was the responsibility of Parliament not the Lords Lieutenant.
- The obligation to serve when called out by the Lord Lieutenant was a legal duty which fell on all citizens between certain ages, originally sixteen to sixty. However, this has varied over the years and eighteen to forty-five has been

mentioned but by 1794 the age bracket was between seventeen and fifty-five and it has broadly settled at this age group since.

- It was the task of the Lord Lieutenant and his Deputies, when so ordered by the Monarch, to draw up lists of those eligible for service in the county and call them out.

- The nature of the obligation differed depending on the wealth of the individual. The poor were simply required to present themselves for service. Those with some property and, or, wealth, at one time judged by the quality of dress of the individual's wife, were obliged to fund their own equipment up to and including the obligation to provide their own horse. Those in this latter category have been described during various periods of history as 'house carls', 'knights' or 'yeomanry'. Perhaps the very first were the Knights of the Round Table.

- The very wealthy and the largest landowners were in addition obliged to raise and to some extent equip whole bodies of men.

- Wealth however also brought privileges including being able to avoid service oneself by nominating an individual to serve in one's place or to choose how one served.

- Those serving were sometimes but not always paid some form of subsistence whilst on duty. Originally called the levy, the meaning of this word evolved to describe the entire process of mobilization which became known as 'the levy' or 'the levies'.

- The organization formed was initially called 'the fyrd' but later became generally known as 'the militia', a title which remained until a reorganization which took place in 1928 sometime after the formation of the Territorial Army which replaced it by virtue of the Haldane reforms of 1908.

- Militia is a name dating from the late sixteenth century and is used to describe citizen-based volunteer forces as opposed to professional soldiers.

- The need to mobilize volunteers arose no fewer than three times in the eighteenth century: firstly to deal with the Jacobite Rebellion of 1745; secondly in the 1770s when the mishandling of the troubles in America resulted in the standing army being found, not for the last time, to be of insufficient size for the commitments the Government of the day had assigned to it and needing reinforcement by the militia; and thirdly as the threat from a combination of republicanism and Napoleon manifested itself at the beginning of the 1790s, by which time, with the benefit of the lessons learnt from the previous two experiences, the process of raising the militia in the, then, modern era had been placed on a much more efficient footing. As a result there existed a capability to raise in excess of 30,000 men in England and Wales. All of this was implemented through the offices of the Lords Lieutenant and their Deputy Lieutenants.

There are several examples of volunteer cavalry regiments being raised in Nottinghamshire broadly by using this process prior to it being used to raise the Regiment. In 1642, Henry Ireton of Attenborough raised a troop which fought for the Roundheads at Edgehill. The next year Colonel Francis Thornhagh of Fenton and Sturton, whose father had undertaken to raise a regiment of Nottinghamshire Horse, took command of this regiment with Ireton as his second-in-command.

In a battle at Gainsborough, Thornhagh was captured but escaped. At Newark he was wounded severely but recovered. Meanwhile, Ireton caught Cromwell's eye and ear and was selected for promotion eventually becoming one of Cromwell's foremost commanders in Ireland.

To illustrate how the Civil War divided neighbours, Robert Pierpont of Holme Pierpont, after some initial indecision, had declared himself for King Charles, who appointed him as his Lieutenant General, and in 1643 agreed to take an armed force to Gainsborough. In the same engagement during which Thornhagh was captured, he was surprised, and attacked by a force under Lord Willoughby of Parham. He and his force fought bravely but they were defeated and so he surrendered. Three days later he was despatched by pinnace down the Trent to Hull. The vessel was fired on by a battery of the Royalists under Lord Newcastle, yet another neighbour. Robert went up on deck to try to stop the shooting, but was straightway hit by a cannon ball which divided him in two.

Meanwhile Thornhagh's Regiment of Nottinghamshire Horse continued to fight for the Roundheads in the Midlands. It routed the Royalists at Rowton Heath near Chester and, in 1648, went to Wales to join in the siege of Pembroke Castle. In August it moved northwards and was on the right flank at Preston where its commander was mortally wounded.

Nearly a century later, in 1745 during Henry Pelham's premiership, a still more significant force was raised to help in the suppression of the Jacobite Rising led by Charles Edward Stewart, 'Bonnie Prince Charlie', son of the Old Pretender, who induced the Highland clans to join him. As most of the standing army was overseas fighting the French in Flanders the 2nd Duke of Kingston-upon-Hull, was given authority to raise and command a regiment of Nottinghamshire Light Horse. Little did anyone realize at the time that the regiment raised would prove to be the prototype for two seperate and equally important genres of cavalry: firstly, until that regiment was raised all cavalry had been mounted on slow weight-carrying horses whose riders were restricted by armour but thereafter all cavalry was mounted on more agile horses ridden by horsemen wearing no personal armour who could thus cover much more ground in a day and attack with greater speed and penetration; secondly it was the forerunner of volunteer cavalry regiments and the yeomanry.

This regiment was raised on 1 October 1745 by the formation of an association which was a process undertaken by a group of private individuals rather than by the Lord Lieutenant, although he would undoubtedly have overseen the process. These individuals pledged the money to raise the regiment by signing an agreement to do so. This process was probably used because the legislation under which such a unit could be raised by the Lord Lieutenant was at the time flawed.

Signatures were obtained pledging a total of over £8,000 towards the cost of the new regiment. The principal subscribers were:

Duke of Newcastle	£1,000
Duke of Kingston	£1,000
Baron Middleton	£400
Duke of Norfolk	£210
Lord Byron	£210
Earl Fitzwilliam	£200
John Mordaunt	£200
Lord Robert Manners–Sutton	£200
Lord Howe	£200
Lord Charles Cavendish	£200
Countess Oxford	£200

The regiment was recruited in Nottingham from 'butchers, bakers, chandlers, and bored apprentices; they were mounted on black and bay horses, with guidons of crimson silk, silver, gold, and green; their short carbines were slung by the ring, tricorne beavers pulled down over their brows, and yard long sabres slapped their great boots. Not a rider was over 5 feet 8 inches, not a horse over 15 hands.'

By the following month the Light Horse had been trained hastily and was in the field near Lichfield in the Duke of Cumberland's Army. On 2 December, it fought its first skirmish near Congleton in Cheshire. Two days later, the Jacobites entered Derby and Cumberland's Army moved to Northampton to bar the enemy from moving on London.

However, on 6 December, the Jacobites withdrew northwards hotly pursued by the 2nd Duke of Kingston and his Light Horse, in the van of Cumberland's advance, through Uttoxeter, Macclesfield, Lancashire and Penrith to Carlisle which fell on 30 December. Several skirmishes were fought, the one at Clifton Moor, just south of Penrith, being of particular note. John Mordaunt, Lord Robert Manners and Lord Byron were amongst the Regiment's officers in this advance of over 200 miles in twenty days.

Early the next year, the Light Horse escorted the Duke of Cumberland through Edinburgh and Perth to Aberdeen and was in a brush with the rebels at Strathbogie on 12 March 1746. On 16 April 1746, the Nottinghamshire men were on the right flank at Culloden Moor where they helped to smash the Highland line and in the pursuit did great execution.

The Duke of Cumberland's command at Culloden consisted of sixteen infantry battalions, all but one regular, three cavalry regiments, of which Kingston's Light Horse was the only volunteer regiment, and a regiment of artillery and so the regiment's contribution was more significant than might be expected of non-regular troops.

On 27 July 1746 the regiment left Fort Augustus and was disbanded at Nottingham on 15 September. Each trooper was given three guineas, a printed copy of the Secretary of War's letter to the 2nd Duke, and was allowed to retain his saddle and bridle.

At the same time the men were given the option of re-enlisting into a regiment of dragoons. All but eight accepted and thus sprang into being the 15th Regiment of Light Dragoons with the Duke of Cumberland as its Colonel and Lord Robert Manners, later Lord Robert Manners-Sutton of Averham and Kelham, as its Lieutenant Colonel. Two years later it charged at the battle of Lauffeld, near Maestricht, in which the British bore the brunt with a firmness that exhorted the praise even of the Frenchmen. It is not often in our history that British cavalry have been in receipt of praise from anyone, least of all the enemy.

The following year peace was declared, economies became the order of the day and the 15th Light Dragoons were disbanded.

In the summer of 1746, Lord Robert Manners-Sutton, Major Chiverton Hartopp, Captain Charles Hatt, Cornet Thomas Smith and Cornet William Hatt were each given the freedom of the Corporation of Nottingham. All that now remains of this Regiment which did so much in so short a time are the relics of two drums lying in safe keeping at the office of the Clerk to the County Justices of Nottingham.

Six families at least who form part of this account were also destined to play a leading role in the history of the Sherwood Rangers. The Pelhams, kinsmen of the Countess of Oxford, who, in the guise of the Dukes of Newcastle, are involved heavily with the Regiment for the first century and more of its life; the Pierreponts, descendants of the Duke of Kingston-upon-Hull who raised and maintained Troops throughout the period up to 1826 when the Nottinghamshire Yeomanry formed into two separate regiments and then continued their association with the South Nottinghamshire Yeomanry; the family of Manners-Sutton of Kelham, a member of which, John Manners-Sutton, commanded the Sherwood Rangers for a period during the mid-nineteenth century; the Barons Middleton who raised troops and were involved closely throughout the first fifty years of the regiment; and the Dukes of Norfolk who were also involved from time to time. Last but not least the Countess of Oxford also represents the Dukes of Portland who were key supporters as well.

In summary it is not unfair to the regular Army (without which the Territorial Army, of which the yeomanry forms part, could not survive today, and vice versa) to make the point that the traditions of the volunteer army are some 900 years older than those of the Regular Army. To illustrate the point, The Royal Monmouthshire Royal Engineers (Militia) and the Honourable Artillery Company, the oldest regiments in the British Army, are volunteer regiments.

As will be seen, the way the ancient process of raising volunteers functioned at the end of the eighteenth century when the Nottinghamshire Yeomanry was raised was firmly based on the long-standing methods described above, and were the same throughout the land. A number of the estate-owning families in a county agreed to raise, equip, and command a Troop of about sixty men. Two or more families often combined for this purpose. The Troops once raised, say half a dozen in number, tended then to be placed under the overall command of a single individual in each county to act as the Commanding Officer. This fairly loose confederation of sub-units was then given a common title and thus a regiment was formed.

What changed with the raising of the yeomanries was that for the first time a group of volunteer regiments, raised and mobilized to meet a specific threat were not disbanded when the threat had been overcome but were maintained afterwards in an unmobilized state as an inexpensive permanent semi-trained resource, principally for home defence, thus capable of responding more quickly and effectively to the unexpected. It is a strategy which has proved itself repeatedly since and has, as a result, been expanded over the centuries into what is now known as the Territorial Army. The birth date of that change however was 1794 and the birth of the yeomanries was the event.

Chapter 2

The Dukes and the 'Merry Men' of Nottinghamshire

Power and Politics

The early years of the eighteenth century, the century in which the Regiment was raised, had seen the end of chairmanship of Cabinet meetings by the King, largely because of the inability of George I to speak English. As a result the role of chairing Cabinet meetings now devolved on one of the members of the Cabinet, whose role was initially more like that of Prince Regent. In time it became the rule that the role fell on the First Secretary to the Treasury, a title still held by the Prime Minister today and thus our modern Parliamentary democracy was born. The name 'Prime Minister' was originally coined by the Opposition as an insult and did not come into official use until the nineteenth century. Until then 'The First Secretary to the Treasury' meant the same.

Sir Robert Walpole emerged in the first half of the eighteenth century and William Pitt the Elder in the second half, as the first true national political leaders. However, these men did not hold office by the popular vote of the people through the ballot box, because, although the ballot box existed, at least for people of significant property, the manner in which people voted was controlled by the nation's most aristocratic and powerful landowning families whose powerbase consisted of being able to exert personal control over the way MPs for a significant number of seats voted, mostly because they had nominated them. The power of these aristocrats had in effect been enhanced by the weakening constitutional position of the King.

The families generally, but not exclusively, were Whigs and the remainder were Tories, although these political affiliations were not nearly as clearly defined as are modern political parties. They are most accurately described as coalitions of power bases based on the various families. Walpole and Pitt, and the other emerging politicians such as North, Addington, Fox and the younger Pitt, held office through the support and patronage of these families.

Of these families, as will be seen, some of the most influential were to be found amongst the four north Nottinghamshire Dukedoms who had caused the area to become known as the Dukeries. The term 'Dukedom' is favoured over that of 'Duchy' because the latter implies both a title and dominion over certain lands going beyond mere ownership which is not the case here.

The Dukes

The four Dukedoms were those of Norfolk, Newcastle-under-Lyne, Portland and Kingston-upon-Hull. Some also include the Duke of Leeds, to make a fifth Dukedom but since all his estates lay just across the border at Kiveton in Yorkshire astride the Sheffield–Worksop road, he is not relevant to this history. It is worth dwelling on him briefly, however, to mention that the 1st Duke of Leeds owes his title to the fact that he acted as William and Mary's 'Prime Minister' during much of their reign.

Of the four Nottinghamshire dukes only one, Norfolk, did not have his principal seat in the county at the time the Regiment was raised. There were many other distinguished families in both north and south Nottinghamshire who were more than influential in raising Troops for the original Regiment, of which the Saviles of Rufford Abbey and the Barons Middleton of Wollaton were the most distinguished. However, so intimately involved were all four of those ducal families with the Regiment over time that some background on each and some understanding of the remarkable power that they wielded in the eighteenth century is essential to understanding the Regiment's ethos and the factors which shaped it.

The Dukes of Norfolk

The Dukedom of Norfolk is the premier non-Royal English dukedom and the Norfolk titles can be traced back to the Bigods. When their line failed their lands reverted to the Crown and Edward II then re-bestowed them on his brother Thomas, whose descendants have frequently played a significant role in the events which make up England's history.

Being so close to the flame of power means that it is impossible to avoid being burnt, and a number of members of the family, down through the centuries, have been slain in battle, beheaded, imprisoned, normally in the Tower of London, or merely stripped of their lands and titles, because they were supporting the wrong Monarch/Pretender at the wrong time. However, being resilient and a large family, it was never long before their titles, prestige and some lands, notably at Arundel in Sussex, were restored to an appropriate survivor.

True wealth was a different matter. For this they relied on one particular stroke of good fortune, namely the marriage of Thomas Howard (1585–1646) to Lady Alethea Talbot youngest daughter of Gilbert Talbot, 7th Earl of Shrewsbury. Although Thomas Howard was the head of the family he was not the Duke of Norfolk, since that title had been stripped from one of his predecessors some time before due to some indiscretion, but he was the 21st Earl of Arundel and the Dukedom would be restored two generations on. Although the 7th Earl of Shrewsbury had two sons, both died young, so on his death the title went to his brother. However, his huge estates in Nottinghamshire, Derbyshire and Yorkshire, including Sheffield, went to his three daughters. Despite Alethea being the youngest, she eventually inherited them all and, through her, so did the Howards. The Howard family's gain was the Talbots' loss.

During the period that embraced the early years of the Regiment's existence the Dukes of Norfolk were, as usual, politically active, the heirs sitting in Parliament for a period as Whig MPs. They enjoy the hereditary office of Earl Marshal of England, the titular commander of the Armed Forces immediately under the Sovereign. They were also Roman Catholics, which makes their survival all the more remarkable.

Even in more civilized times, survival has not always been smooth. Indeed Charles, the 11th Duke (1746–1815) who succeeded to the title in 1786 and therefore held the title when the Regiment was formed, had renounced the Catholic faith in 1780 (no doubt with his fingers firmly crossed). This was at the time of the Lord George Gordon riots which were staged in opposition to the partial emancipation of Roman Catholics. He made the renunciation in order to retain influence. Not much good did it do him, however, since in 1798, in accordance with family tradition, he managed to get himself deprived of all his offices by George III, including that of Lord Lieutenant of the West Riding of Yorkshire. His offence was a toast, regarded by the King, a supporter of the Tories, as rather too radical for his liking, that he had given at a political dinner attended by approximately 2,000 people. The toast was *'Our Sovereign's health – the Majesty of the People'*. Ironically the news of his demotion reached the Duke when he was entertaining the Prince of Wales, the future George IV, to dinner. There was something of a Hanoverian tradition that members of the Royal Family were heavily involved in politics, the heir almost always on the opposite side to the Monarch.

Worksop Manor was the Dukes of Norfolk's Nottinghamshire estate, part of their inheritance from the Earls of Shrewsbury, although, at 6,000 acres it was much smaller than the 20,000 acres they held in West Yorkshire, principally in and around Sheffield, and at Arundel in Sussex both of which over the years were more their principal residence than was Worksop Manor. Although the acreage with Worksop Manor was relatively modest compared with the other great estates of the Dukeries the historical importance of the house itself and those which preceded it on the site was unsurpassed in the north of the county and derived from its role as a Royal Hunting Lodge.

As can be appreciated from the foregoing, the Dukes of Norfolk played a lesser role than some, in the affairs of both the Regiment and the county, but, as will be seen, the family became closely involved with the Regiment for a time.

Bess of Hardwick

Elizabeth Talbot, Countess of Shrewsbury, (1527–1608), was the third daughter of John Hardwick of Hardwick Hall in Derbyshire. This remarkable woman exerted a crucial influence over each of the three remaining ducal families and, for good measure, over the Saviles as well. She married four times, accumulating in the process vast estates, the last marriage being to George Talbot, 6th Earl of Shrewsbury. For the purpose of this narrative, the second marriage to Sir William Cavendish was the important one. They had eight children, of whom three are worth noting. William Cavendish (1551–1626) was the forebear of the Dukes of Devonshire, who inherited Hardwick Hall, Derbyshire; Welbeck Abbey, Nottinghamshire and the Cavendish estates based upon Chatsworth, Derbyshire; Charles Cavendish (1553–1617) was the forebear of, and responsible for a significant part of the inheritance of, both the Dukes of Newcastle and the Dukes of Portland. He inherited lands in Derbyshire and Nottinghamshire, including Nottingham Castle, and Bolsover Castle, Derbyshire. Lastly there was Frances, one of their daughters, who married Robert Pierrepont the forebear of the Dukes of Kingston-upon-Hull.

Charles Cavendish's son became ennobled as the 1st Duke of Newcastle. He lived at Bolsover Castle which he restored and enhanced into a spectacular building. It was this fact which inspired his title of 'new castle'. The Duke was renowned as a great classical equestrian which, in those days, was based on the skill of imparting the elaborate dressage movements in which a war horse had to be trained to enable a nobleman to fight from its back using a sword. From this came the use of the word 'Cavaliers' to describe the supporters of the Stuart dynasty and the flamboyant dress they wore.

His son, the 2nd Duke had no male heirs but had a daughter, Lady Margaret Cavendish (1661-1716), who married one John Holles, whose ancestors originated from London where they had made a great fortune and land holdings by lending to landowners on the security of their estates who then defaulted. He, in turn, was ennobled with the title 1st (for the second time) Duke of Newcastle-upon-Tyne. Avaricious bankers are not a new phenomenon. They did not have a surviving male heir either. However, they did have a daughter, Lady Henrietta Cavendish Holles (1694-1755). For some reason the new 1st Duke did not want her to inherit, preferring instead the son of his sister. The son's name was Thomas Pelham. This decision was contested and, as a result, a majority of the estate was inherited by Lady Henrietta, including Bolsover Castle and the remainder, including Nottingham Castle, went to Thomas Pelham, who changed his name to Thomas Pelham-Holles in 1711.

Meanwhile Lady Henrietta Cavendish Holles married Edward Harley 2nd Earl of Oxford and their daughter and heiress Lady Margaret Cavendish Harley married William Bentinck 2nd Duke of Portland bringing her inheritance with her. Naturally the title of Duke of Newcastle-upon-Tyne, in due course, died out for the second time.

The Dukes of Newcastle-under-Lyne
There were two titles. Thomas Pelham-Holles already mentioned (1693–1768), who was created the 1st Duke of Newcastle-under-Lyme (later changed to Lyne) in 1756, was already the 1st (for the third time) Duke of Newcastle-upon-Tyne. He had also inherited the barony of Pelham of Laughton from his father in 1712. By the time he came of age in 1714, he had become one of the wealthiest Whig landowners in England, with holdings in twelve counties and a rental income of nearly £40,000 a year (multiply by 150 to compare with today's values), so although he apparently lost out to his cousin Lady Henrietta Cavendish, he had not done too badly.

The reason the second title was created was because the Duke had no heirs and therefore the first title would become extinct, for the third time, on his death. The second title was created with a special remainder in favour of the Duke's nephew Henry Fiennes Clinton, 9th Earl of Lincoln, and this overcame the problem.

Thomas Pelham-Holles, Duke of Newcastle, must have been a very powerful man to be granted a second Dukedom when he had already been granted a perfectly good one earlier. He was indeed one of the most powerful in the land. He had been created Duke of Newcastle-upon-Tyne in 1715 as a reward for helping to bring about the succession of King George I to the throne (reigned 1714–27). Not only that, but the Duke and his younger brother the Right Honourable Henry Pelham, initially in alliance with Sir Robert Walpole, and after his retirement in their own right, were no less than the leading politicians of the first half of the eighteenth century. The two brothers worked as a team, with Henry the more able but it was the Duke who had developed the greatest powerbase. Initially it kept Sir Robert Walpole in office as the first 'Prime Minister' for twenty-one years from 1721 to 1742 – not only the first, but the longest premiership in history. It then kept Henry Pelham in office from 1742 until his death in 1754. Finally it kept the Duke himself in office from 1754 to 1762, save for a period when, having precipitated the Seven Years War, he had to resign briefly in favour of his kinsman the Duke of Devonshire. When he resumed office, however, it was in a coalition which signalled the end of the Whig supremacy. The fact that the Liberal Party is the natural political successor of the Whig tradition hints of history repeating itself. Throughout the entire period of the premiership of the other two, the Duke held high Cabinet office.

Churchill wrote of him, 'Newcastle, in his own whimsical way, looked upon the work of government as the duty of his class, but he had no clear ideas as to how to discharge it'. Lord Shelburne, himself appointed 'Prime Minister' in 1782, said of the two brothers,

They have every talent for obtaining Ministry, and none for governing the kingdom, except decency, integrity, and Whig principles … Their forte was cunning, plausibility and the cultivation of mankind; they knew all the allures of the court; they were in the habits of administration; they had long been keeping a party together … Mr Pelham had a still more plausible manner than his brother, who rather cajoled than imposed upon mankind, passing for a man of less understanding than he was.

If that was so then he succeeded in convincing George II who referred to the Duke as 'the impertinent fool'. The fact is, however, that no one controls the greatest nation on earth for forty-one years without formidable political skills of a sort, what is more, which seem to resonate no less clearly at the beginning of the twenty-first century.

When the 1st Duke died in 1768 the title duly passed to Henry Fiennes Pelham-Clinton, 9th Earl of Lincoln (1720–1794). He was forty-eight years old and had inherited his earldom aged ten; the 1st Duke was his guardian. That is why he came to regard Lord Lincoln as his heir. The Clintons traced their lineage back to the conquest, an ancestor having fought with William the Conqueror, for which he was awarded the Lordship of Clinton in 1067.

After university, Lord Lincoln went abroad to Turin to study fencing. Horace Walpole, the writer and son of Robert Walpole, and Thomas Gray the poet who famously wrote 'Gray's Elegy' had both been friends of his at Eton and were also there. Horace Walpole dropped his friendship with Thomas Gray, in favour of Lord Lincoln, and then in turn quarrelled with him. Lord Lincoln was exceedingly good-looking, and would later have the reputation as the most handsome man in England.

Not only did he inherit the title of the 2nd Duke of Newcastle but he also inherited much of the 1st Duke's political powerbase at both national and local levels. Clumber Park in north Nottinghamshire became his principal estate. He had adopted the surname Pelham because he had married his cousin Catherine Pelham, Henry Pelham's daughter and so was the true inheritor of all the influence that the Pelham brothers, his uncles, had built up.

At various points in his life the 2nd Duke became a Privy Councillor, Cofferer of the Household, Comptroller of the Customs for the Port of London (a very lucrative appointment), and Knight of the Garter but, unlike his uncles, he was not so much interested in the exercise of power as the exercise of influence over power at which he was most adept and used freely over the leading politicians of the day, thus sustaining the family's leadership nationally. He achieved this by exercising political influence in eight parliamentary seats, including that of Westminster where he was High Steward.

One example of his power was that King George III turned to him to provide the support of the six MPs whom William Pitt the Younger needed to secure the majority with which to form a government in 1783. Although the 2nd Duke did not pursue a political career nationally, he was active in local elections. He also used his influence to promote the career of his cousin Sir Henry Clinton, a career army officer. Thanks to the Duke's persuasion, Sir Henry was appointed commander-in-chief of the British forces in America during the American War of Independence. As part of the quid pro quo, Sir Henry took the Duke's son and heir, Thomas, as his aide-de-camp.

In addition he was made Lord Lieutenant of the counties of Cambridgeshire in 1742 and Nottinghamshire in 1768, the latter in succession to the 1st Duke. Both of these offices he held until his death in February 1794, just before the legislation through which the yeomanry, and therefore the Regiment, was raised became law. Horace Walpole described him in 1758 as the 'adopted heir to the Duke of Newcastle and the mimic of his fulsome fondness and follies, but with more honour and more pride'.

Thomas, the 3rd Duke of Newcastle (1752–1795) succeeded to the title on the death of his father in February 1794. He was the third son of the 2nd Duke, his older brothers having predeceased him. Following his tour as Sir Henry Clinton's ADC he had enjoyed a successful military career, which included a tour as ADC to George II, retiring as a major general in 1787 aged thirty-five, by which time he had already developed a second career as a politician. He sat as a Tory MP for Westminster from 1774 to 1780 and East Retford from 1779 to 1794 (note the change in his political alliance, which very much reflected the majority political position of the day), giving up his seat in the House on succeeding to the title. On the death of his father he was appointed Lord Lieutenant of Nottinghamshire and therefore, although he died unexpectedly in 1795 aged only forty-two, due to an adverse reaction to a 'cure' he had taken for whooping cough, it was he, as Lord Lieutenant, who implemented the formation of the Regiment.

It is understood that because the 2nd Duke disapproved of the 3rd Duke's secret marriage to Anna Maria, the youngest daughter of the Earl of Harrington, he left all his estate to the 3rd Duke's eldest son leaving the 3rd Duke's income, according to Lady Stafford, 'so small as there is hardly enough to keep him out of Debt'.

On his death he was succeeded by the aforementioned son Henry Pelham-Clinton, 4th Duke of Newcastle (1785–1851), who, as has been mentioned, had already been endowed with all the material things of life that he could possibly want by his grandfather and who was destined to command the Regiment but was, at the date of his succession, only ten years old. The 3rd Duke was succeeded as Lord Lieutenant by the 3rd Duke of Portland.

The Dukes of Newcastle's land holdings were upwards of 35,000 acres, the vast majority of it in Nottinghamshire.

The Dukes of Portland
The Cavendish-Bentincks, Dukes of Portland, trace their line through Hans Willem Bentinck (1649–1709), who accompanied William of Orange from the Netherlands and, in recognition of his services, was made Earl of Portland in 1689 when the Prince of Orange became King William III. He was granted a large number of estates which were added to by following generations, often through advantageous marriages, and his son, Henry (1682–1726), was created 1st Duke of Portland in 1716.

As has been mentioned one of those advantageous marriages was by the 2nd Duke to Lady Margaret Cavendish Harley. They promptly recognized the Cavendish family in their own surname. She brought with her Welbeck Abbey and Bolsover Castle and the large landholdings which went with them.

It was their son, the 3rd Duke of Portland (1738–1809), who was the incumbent when the Regiment was raised. He had married Dorothy Cavendish, a daughter of William Cavendish, 4th Duke of Devonshire. He was one of the, (if not the most), distinguished, members of the family and had succeeded to the title in 1762. One of the leading Whig politicians of his age, he began his political career by sitting briefly as MP for Weobley, Herefordshire, from 1761 to1762, resigning his seat on his father's death. He held Cabinet office more or less permanently from 1765 until his death. He was First Secretary to the Treasury ('Prime Minister') not once but twice, the first time briefly in 1783, which was the Ministry formed by the coalition of North and Fox which had to give way to that of Pitt the Younger due, it should be remembered, to the support given to Pitt by the un-neighbourly 2nd Duke of Newcastle, and, secondly, between 1807 and 1809 during the early period of the Napoleonic Wars. He died shortly after leaving office. In between he was Home Secretary between 1794 and 1801 which was regarded as his most successful period of office, and must have included a leading role in the introduction and implementation of the legislation enabling the formation of the yeomanry and thus the Regiment. He was also Lord President of the Council between 1801 and 1805 in both Addington's and Pitt's Governments. Among many other offices he was, as has been mentioned, Lord Lieutenant of Nottinghamshire from 1795 until his death in 1809 and was described by contemporaries thus: 'without any apparent brilliancy his understanding is sound and direct, his principles most honourable, and his intentions excellent'.

The Dukedom's Welbeck Abbey Estate in north Nottinghamshire which ran to over 43,000 acres was now one of its principal residences. The estate was by no means the totality of the family's holdings which consisted of another 20,000 acres elsewhere in England and over 118,000 acres in Scotland, 181,000 acres in all. Together the Dukedom's holdings were among the ten largest individual land holdings in the country.

Although the 3rd Duke did not himself serve in the Regiment a number of his successors did.

The Dukes of Kingston-upon-Hull

This title was that of the Pierrepont family who have been based in Nottinghamshire for many centuries. They are descended from Robert de Perpont who came to England with the Normans and had become both wealthy and large landowners by a combination of shrewd marriages, one to the daughter of one of Henry VII's tax collectors, another to Annora Manvers of Holme, one to a granddaughter of the Earl of Shrewsbury and yet a fourth, as already mentioned, to Frances, the daughter of Bess of Hardwick from her marriage to Sir William Cavendish. They had also made shrewd purchases of land, including monastic lands from the Dissolution of Chantries in the reign of Edward VI.

Robert Pierrepont, the son in law of Bess of Hardwick, bought himself the titles of Viscount Newark and Earl of Kingston-upon-Hull. Their principal seat was at Holme Pierrepont on the bank of the Trent which had always been a substantial property and which they had enhanced over the centuries, but they also owned another large estate at Thoresby in North Nottinghamshire. Robert Pierrepont's great grandson, Evelyn Pierrepont (1655–1726) entered politics, becoming the Member for East Retford and was created Marquess of Dorchester for his involvement in the arrangements for the Union with Scotland. He later became involved politically in the business of the House of Lords and in 1715 was created Duke of Kingston-upon-Hull. All these titles were inherited in due course by his grandson Evelyn Pierrepont (1711–1773), who became the 2nd (and last) Duke of Kingston. He married Elizabeth Chudleigh in 1769, but had no male heir and the titles died out on his death, although he did have a sister who bore two sons.

The 2nd Duke's sister, the Lady Frances, had married the Deputy Ranger of Richmond Park, one Philip Medows. He was said to be worth not more than £900 whereas she was worth £20,000 a year; further she had showed great style by declining a settlement of £16,000 per annum and the sum of £100,000 offered by the Duchess of Marlborough 'provided she married Jack Spencer, her Grace's grandson'. Contemporary opinion concluded that, on the basis of this evidence, Lady Frances' marriage to Philip Medows must have been a love match, an event not entirely approved of in polite society at that time. The Duchess of Marlborough's opinion on the matter can be imagined, although it was of no interest whatsoever to Lady Frances as she happily set up home as Mrs Philip Medows in 'Ranger's Lodge' in Richmond Park. What a good name for a house.

Lady Frances had two sons, Evelyn, who was the elder, and Charles, but the 2nd Duke, having left his wife the Duchess of Kingston all his estate's income for her life, named the second son, Charles Medows (1737–1816) to be the heir to his estates in Nottinghamshire.

Evelyn challenged the will and commenced proceedings which eventually found the Duchess, his aunt, to have already been married to a man called

Captain Hervey when she married the Duke. The challenge led to the Duchess being charged with bigamy. As a Peer of the Realm she claimed her right to impeachment by her fellow Peers in the House of Lords. The Duke of Newcastle was a close friend and neighbour and provided support. The Peers concluded that the will was sound, despite her bigamous marriage which was proved, but coincidently the House of Lords went into recess before concluding their sentence, so allowing the Duchess to depart into self-imposed exile to her estate on the Polish/Russian border. She died in 1788.

Charles Medows joined the Royal Navy in 1755 at the age of eighteen and was almost immediately promoted to lieutenant and then post captain at the incredibly early age of twenty. He is credited by Horace Walpole with taking the *Hermione*, a Spanish galleon containing a rich consignment of bullion, whilst Captain of the *Isis*. In 1757 he was promoted to command the sloop *Renown*, and then the 36-gun frigate *Shannon*, taking part in one of the earliest combined military/naval operations which led to the capture of Louisbourg, thus preparing the way for the fall of Quebec and establishment of British rule. He was entrusted with the task of conveying the messengers carrying the news of the French capitulation in July 1758 to the King. He was then twenty-one. The following year he was promoted to the rank of captain, continuing at this rank for the next ten years before retiring in 1773 when the Duke died.

The Medows had taken up residence in Nottinghamshire following the 2nd Duke's death, and had opted, as had the Duke, to live at Thoresby Park, which had been built relatively recently rather than at Holme Pierrepont Hall. On leaving the Navy Charles Medows was sponsored by the 2nd Duke of Newcastle to represent Nottinghamshire in Parliament as a Whig from 1778 to 1796. He was a supporter of the 3rd Duke of Portland. It was during this period that he inherited Thoresby and Holme Pierrepont on the death of the Duchess in 1788, at which point he also assumed the surname and arms of Pierrepont.

In 1796 with the support of the 3rd Duke of Portland he was created Baron Pierrepont of Holme Pierrepont and Viscount Newark of Newark-on-Trent and gave up his seat as an MP. Furthermore, in 1806, he had conferred on him, by the same means, the title of Earl Manvers. This was a new title, and apparently the reason why he did not choose to re-create the name Kingston-upon-Hull was because he preferred to acknowledge the previously-mentioned marriage of Annora Manvers of Holme to Henry Pierrepont in the thirteenth century and therefore her place as the most important ancestress to the family and its estates in Nottinghamshire. It was he, therefore, who was the incumbent at Thoresby when the Regiment was raised. His estates were thought to consist of approximately 38,000 acres of which some 27,000 lay in Nottinghamshire. His circumstances were a far cry from Ranger's Lodge, Richmond Park, where he was brought up.

The total acreage controlled by these four families was about 300,000 acres with approximately 110,000 in Nottinghamshire, the majority of which marched together. It has to be added that up until the Middle Ages a significant part of the land concerned formed part of the Royal Hunting Forest of Sherwood, which was much valued by the monarch of the day. Since several of the predecessors of these distinguished families between them held the Stewardship of the Forest on trust from the Crown, and there is little evidence of any Monarch making material divestments of the land, it is left to speculation as to how the land became owned by them. It is said that the explanation is that the sporting rights and the ownership were held separately.

Be that as it may in the eighteenth century this sort of acreage, by then held beyond dispute, was real power. Of the four, undoubtedly the most powerful family, if not the most senior, both nationally and in Nottinghamshire, was that of the Dukes of Newcastle.

The 'Merry Men' of Nottinghamshire

The Dukes had power, but power is nothing without influence and, although they had some of that too, history shows it was insignificant locally compared with that wielded by the people of Nottinghamshire. It is no exaggeration to say that the Sherwood Rangers Yeomanry owes its very existence and survival far more to the people of the county than to the people who actually raised it, which is why this book is dedicated to the former. This truth is confirmed from whatever angle it is examined. On the face of it Nottinghamshire folk are typically good citizens – hard working and responsible – but of a good humoured, companionable and notably un-rebellious nature.

However appearances are deceptive. The plain fact is that the Regiment, once raised, was retained in service long after the vast majority of similar units in other parts of the country had been disbanded since in few counties were the causes of our social revolution (for example, the issues concerning republicanism, the Corn Laws, Luddism, parliamentary reform, and Chartism) more keenly promoted and championed by the emerging working class than in Nottinghamshire, particularly in the first half or so of the nineteenth century. This may in no small measure have been due to the fact that, at that time, Nottingham was considered to have some of the worst slums of any city in the country and that this was caused by the fast growing town being surrounded by large estates whose owners refused to allow land to be made available to accommodate that expansion.

The 'Merry Men' carried greater influence than the Dukes because the Dukes opposed them, calling out the Regiment to maintain law and order, but in the vast majority of cases the 'Merry Men' eventually prevailed and their causes became established and now form part of the central core of our democracy and our way of life. In that way the nation owes a considerable debt to those parts of the country, of which Nottinghamshire was one, where these issues were raised most vigorously and fought for and to which the Regiment owes its survival during that time when most others were disbanded.

There is another side to this because the emerging working class did not always get it right, for example republicanism and Luddism were causes felt as keenly in Nottinghamshire as anywhere in the land and as keenly fought for by the working class, but in respect of which history has found against them. In these cases the Nottinghamshire folk we thank are those, including the Dukes who espoused not only Loyalty to the Crown, but also championed and gave birth, through their entrepreneurialism and vision, to the Industrial Revolution and advanced agriculture and fought for these changes against the determined reactionaries. The only logical conclusion to draw from the level of this type of activity experienced in Nottinghamshire compared with many other places is that there must lie, normally dormant, beneath the people of Nottinghamshire's typically even-tempered exterior, a much more resolute, one might almost say implacable, trait.

Finally there are the 'Merry Men' who have served in the Regiment down the years, initially those who clattered about the county on their horses, as their cows remained untended, maintaining the King's Peace and, later, those who fought with quite exceptional, one might say implacable, resolution during the wars of the twentieth century and, finally, those who today give of their free time in service of their country in places as far afield as Iraq, the Balkans, and Afghanistan.

Chapter 3

The Raising of the Regiment 1794

'The pith and substance of this country'

The Regiment would never have been raised if Marie Antoinette, Queen to Louis XVI of France, had not suggested to the French peasantry that they should supplement a diet seriously deficient in bread by eating cake. Although this statement is entirely apocryphal, the French Revolution, in the popular mind, has often been attributed to her alleged remark. Nonetheless, the Revolution, which erupted in 1789, was the reason for the raising of the Nottinghamshire Yeomanry from which the Sherwood Rangers Yeomanry is descended.

Revolution came about in France due to a dangerous cocktail of lack of democracy, unfair taxation, low incomes and high prices, particularly the price of wheat due to poor harvests due to exceptional weather conditions said to be triggered by volcanic activity in Iceland which affected at least two harvests at that time (the pretext for the Queen's remark), to which was added land ownership, which was largely concentrated in aristocratic hands. These were exactly the same factors as existed in Britain and elsewhere in Europe, and so developments in France were watched keenly and with growing alarm by the ruling aristocracy and with eager anticipation by potentially rebellious republicans everywhere else.

In France there were, however, two additional factors which were not to be found elsewhere and which in the end singled her out from other nations. The first was that for many years the Kings of France had kept rebellion at bay by ensuring that all the aristocracy were permanently resident at the King's Court, latterly at Versailles, just outside Paris, where he could keep an eye on them. This meant that they were absentee landlords from their estates and were, therefore, not well known to their own estate workers and, in turn, did not know them. These landowners, therefore, had little knowledge or understanding of their estate workers' poor circumstances and were unsurprisingly in return perceived as little more than parasites by these people, who felt no loyalty towards them.

The second was the apparent passion of the French, whether in the blood or born of the Revolution, for 'manning barricades' – in other words for flamboyant gestures of defiance

At first, perhaps surprisingly, the Revolution, led by Robespierre, proceeded on political rather than rebellious lines. Steps were negotiated to adjust the political system to share power and to redress inequalities. When it became apparent that there was the inevitable and significant gap between what the King and the aristocracy were prepared to concede compared with where the peasants and the middle classes, known as the *bourgeoisie*, wished to end up, the process ground to a halt, and unrest began to build. In 1792 the King attempted to flee the country with his family. They were following many from the upper classes known as the *émigrés* who had preceded them.

Unfortunately for the King he was apprehended in the process and escorted back to Paris, where, on 21 January 1793, to the horror and outrage of the rest of Europe he was hauled before the mob and guillotined. Worse still, Europe then looked on aghast as, at the rate of up to fifty per day, a wholesale and relentless slaughter of the aristocracy commenced, through a process beginning with arrest and imprisonment followed by peremptory trials and ending with a ride in a tumbrel to the same grizzly, but locally popular, method of execution as befell the King in front of the ever present mob.

The Prussians and Austrians, aided by an army of French *émigrés*, reacted first. In the autumn of 1793 they invaded France where they were unexpectedly but nevertheless roundly defeated at Valmey by the infinitely weaker and poorly equipped French under Dumouriez, and then chased back out of France. The French continued the pursuit until, in January 1794, they had taken the Austrian Provinces of the Netherlands, including the port of Antwerp. The French now had 30,000 troops under arms in Antwerp pointed at England. Although such a small force, unsupported by a navy, could not possibly have sustained an invasion, they were well aware that there existed a republican movement in England. Therefore in order to exert maximum pressure and in the hope of sparking a spontaneous revolution in England, they now declared War on Great Britain and Holland. It should be added, that their antagonism towards England may have been fuelled by the belief, whether justified or not, that Britain had been instrumental in causing civil unrest in the Vendée in western France.

On 5 March 1794 the Prime Minister, William Pitt, the Younger, stated 'We are at War with those who would destroy the whole fabric of our Constitution'. He proposed to Parliament measures for strengthening both the nation's defences and the arrangements for internal security, which included the raising of volunteer units. These proposals consisted of various provisions to raise and increase cavalry, infantry, and pioneer units by the long-established process of imposing on all civilians of military age a compulsory obligation to serve in one or the other. The cavalry units so raised were called 'fencible cavalry' and many such units were raised in the ensuing emergency.

As part of the plan for volunteer units, Pitt went on to say,

> As an augmentation of the cavalry for internal defence was a very natural object, they might under certain circumstances have a species of cavalry consisting of Gentleman and Yeomanry, who would not be called upon to act out of their respective counties except on the pressure of an emergency, or in cases of urgent necessity.

In due course the enabling legislation was published and paragraphs v and vi referred to this new idea of yeomanry cavalry:

> v.   Bodies of cavalry to be raised within particular districts or counties, to consist of Gentlemen and Yeomanry or such persons as they shall recommend, according to plans to be submitted to and approved by the King or Lords Lieutenants of Counties under authority from him.
>
>> The officers to receive commissions from His Majesty and the muster rolls to be approved by the same.
>>
>> No levy money to be given.
>>
>> The members to find their own horses.
>
> vi. The Arms and Accoutrements to be given by the Crown, only to be exercised by Royal Warrant of the Lords Lieutenants; only liable to be called out of the County or embodied by the Lord Lieutenant by Royal Warrant or by the High Sheriff of the County for the suppression of riots or tumults within their own or adjacent counties or by Royal Warrant in case of invasion, in either case to receive pay as Cavalry and to be subject to Military Law.

Note that the emphasis was on the 'suppression of riots or tumults' rather than on 'resisting invasion'.

The new form of cavalry unit was called yeomanry after the wording in the Act. For the individual, the attraction in joining the yeomanry was that, since service of some sort was obligatory, service in a unit which was relatively exclusive and which could not ordinarily be required for service outside the county boundary, let alone overseas, was preferable to service in the militias or the fencible cavalry where there existed an unlimited risk to serve wherever required.

Pen & Sword Books
FREEPOST SF5
47 Church Street
BARNSLEY
South Yorkshire
S70 2BR

2

If posting
from outside
of the UK
please affix
stamp here

DISCOVER MORE ABOUT MILITARY HISTORY

Pen & Sword Books have over 1500 titles in print covering all aspects of military history on land, sea and air. If you would like to receive more information and special offers on your preferred interests from time to time along with our standard catalogue, please complete your areas of interest below and return this card (no stamp required in the UK). Alternatively, register online at www.pen-and-sword.co.uk. Thank you.

PLEASE NOTE: We do not sell data information to any third party companies

Mr/Mrs/Ms/Other.............. Name..

Address...

.... Postcode...............

Email address...
If you wish to receive our email newsletter, please tick here ☐

PLEASE SELECT YOUR AREAS OF INTEREST

Ancient History ☐	Medieval History ☐	English Civil War ☐
Napoleonic ☐	Pre World War One ☐	World War One ☐
World War Two ☐	Post World War Two ☐	Falklands ☐
Aviation ☐	Maritime ☐	Battlefield Guides ☐
Regimental History ☐	Military Reference ☐	Military Biography ☐

Website: www.pen-and-sword.co.uk • Email: enquiries@pen-and-sword.co.uk
Telephone: 01226 734555 • Fax: 01226 734438

The idea of yeomanry type terms of service had originated from two sources. The first had been that of Arthur Young, the famous Suffolk agriculturist, who believed in the concept of a Militia of Property for which he had been trying to win support from the landed classes for the previous two years through a paper he had published under the all too explicit title *The Example of France a Warning to England*. The second was a groundswell of insecurity in south-eastern England, which had grown throughout 1793 due to the worsening situation across the Channel, and was centred on the Cinque Ports of which Pitt was the Lord Warden. There was perceived to be a need for a loyal force of light cavalry, as a counter to the admittedly remote risk of invasion on the basis that although invasion was unlikely if it occurred it would occur on the Kent or Sussex coast. This had been conveyed to the government, which had initially resisted but had, in the new and more dangerous circumstances now prevailing, elected to adopt it.

Shortly after the enactment of this legislation, the 3rd Duke of Newcastle who was, as already mentioned, the newly-appointed Lord Lieutenant of Nottinghamshire, laid plans for one of this new type of cavalry regiment to be formed in the county. So it was that three months later, on 10 June, a general meeting of the County of Nottingham assembled at the Moot Hall at Mansfield, under the chairmanship of The Right Honourable Frederick Montague, and unanimously passed the following resolutions:

> Firstly: That in the present crisis it is very desirable to increase the internal force of the County under the sanction of Parliament, by a voluntary subscription.

> Secondly: That the mode which appears most adapted to the situation and circumstances of the County is that of raising a Corps of Cavalry, composed of Yeomen agreeable to the plan suggested by the Government to the Lords Lieutenant of several counties.

> Thirdly: That a subscription be opened for the above-mentioned purpose, and that subscriptions be received at the different banking houses of the County.

Note the phrase 'increase the internal force of the County' which again shows that the emphasis was on internal security.

Anthony Hardolph Eyre, of Grove, was appointed Lieutenant Colonel of the whole Corps and William Boothby was appointed Major. The meeting on 10 June also addressed the naming of the Regiment and, further, purported to address the question of the uniform and the level of expense to be incurred in equipping each man by passing the following resolutions:

That the several troops to be raised be called the 'the Nottinghamshire Yeomanry', and that the uniform of the whole be a dark green lined with buff, cape (Collar) and cuffs double breasted lined with buff, and buff waistcoats, with leather breeches, and yellow buttons with the words 'Nottinghamshire Yeomanry' round them; each officer to have two epaulets. The cloaks to be green edged with buff. The whole will be according to patterns to be prepared as soon as possible and left with the commanding officer of each troop.

Resolved: That the committee will allow to the commanding officer a sum not exceeding Fourteen Pounds to pay for the several articles of Clothing, Accoutrements and Furniture of each Volunteer.

	£.	s.	d.
A Coat and Waistcoat to the uniform hereinbefore agreed to	3	3	0
A Pair of Boots with Hussar tops	1	4	0
A Hat with Feather, Bearskin and Cockade	1	1	0
Military Bridle, Saddle Furniture and complete and Uniform Spurs	5	3	0
A Pair of Leather Breeches	1	7	0
A Cloak	2	2	0
	14	0	0

Given the clarity of this resolution why was it that there is no record or evidence whatsoever that in its early years the Nottinghamshire Yeomanry wore anything other than 'Scarlet faced with Buff'?

The answer may well lie in the next step taken, which was to implement the decision to open a subscription to raise the considerable sum that would be required to equip the Regiment. The appeal appeared to follow the precedent set by the 2nd Duke of Kingston forty-nine years before. In all, the remarkable sum of £8,549 1s 0d was subscribed, almost exactly the total previously raised by the Duke of Kingston, but worth only two thirds. (As a rough guide multiply by one hundred to relate the sum to today's values.) The leading subscribers were the 3rd Duke of Newcastle, the 3rd Duke of Portland and Baron Middleton of Wollaton Park who each subscribed £300. With the amount raised by the subscription now known, and the feasibility of raising volunteers in various parts of the county assessed, a further meeting of the committee took place at East Retford on 8 July to put the resolutions into effect.

It was decided that the 'Nottinghamshire Yeomanry' would be formed consisting of four troops of not less than sixty nor more than eighty strong in men and officers, one at each of Retford, Mansfield, Newark and Nottingham, about 250 all ranks in total.

The answer to the apparent change in the decision regarding the uniform may well be that when the original resolution was passed it was not known that the subscription would raise as much as it did, which was far more than was required to equip in the manner originally decided a unit of the size now proposed. It is known that the committee did, in addition to the items originally listed, buy sabres for the whole force which at that time cost about £1 10s 0d each. Even so it would take little over half the money raised to equip the force with the strength by then decided upon. This explanation seems to be corroborated in a letter that Lieutenant Colonel Eyre wrote some years later and is quoted more fully in due course, in which he said:

> I own I always was of opinion that as little expense as possible ought to be incurred in equipping Corps of Volunteers except in articles that were really useful: and though our Corps (The Nottinghamshire Yeomanry) has not acted up to this principle yet it is well-known that I only gave way to the wishes of others.... And ... our uniform is unfortunately expensive ... our jacket alone costs five pounds.

There is no explanation for the change from green to scarlet, but since one of the key roles of the Regiment was to act as a high profile warning to potential troublemakers, scarlet is undoubtedly a much more dominant and suitable colour for that purpose than green. Anyone who has ever served in the Regiment will not be able to read the foregoing controversy without experiencing the warm glow of recognition that the tradition whereby yeomen never dress as you wish them to, or as they are supposed to dress, was established on the very date of the Regiment's foundation.

The next stage was to recruit the officers. Early in August the committee met again and nominated the officers as follows:

Retford Troop   Captain Anthony Hardolph Eyre of Grove
                   Lieutenant Samuel Crawley
                   Cornet Robert Ramsden

Mansfield Troop   Captain William Boothby of Edwinstowe
                   Lieutenant John Sutton of Scofton
                   Cornet Thomas Woollaston White of Wallingwells

Newark Troop	Captain John Denison of Ossington (succeeded by Captain Francis Chaplin in November 1794)
	Lieutenant Phillip Palmer of East Bridgford
	Cornet Francis John Brough of Newark
Nottingham Troop	Captain Ichabod Wright of Mapperley
	Lieutenant William Richard Middlemore
	Cornet Alexander Hadden

As can be seen Anthony Hardolph Eyre, as well as being Lieutenant Colonel of the whole Corps was Captain of the Retford Troop and William Boothby commanded the Mansfield Troop as well as being the second in command.

The chairman of the committee transmitted these resolutions, on 9 August 1794, to the Lord Lieutenant, the Duke of Newcastle, and thence to the Duke of Portland, the Home Secretary, who replied on 19 August:

> MY LORD
>   I have the honour to acknowledge receipt of Your Grace's Letter dated 15th instant enclosing the names of Gentlemen recommended to hold commissions in the different Troops of Yeomanry Cavalry now raising in the County of Nottingham, and to inform Your Grace that I have transmitted the same to Lord Amherst in order that he may have his Majesty's pleasure thereupon
> I am etc., etc.
> PORTLAND

The officers were approved and gazetted officially on 15 August. This date is important because many years later it would be used in conjunction with the condition of unbroken service to identify the seniority of the Regiment in relation to other yeomanry regiments. In fact 9 August, the date on which the resolutions forming the Regiment were passed and handed to the Lord Lieutenant, was probably the date that should have been used.

In point of fact the Regiment was by no means one of the earliest to form. The first, unsurprisingly, was the Cinque Ports Yeomanry Cavalry on 23 March, and by 15 August, when the Regiment formed, a total of thirty-six others had already done so. How the Regiment eventually became fourth in seniority is a gem of an example of military decision-making that will be saved until later. Given the close involvement of the 3rd Duke of Portland in the Government's plans in his capacity initially as Home Secretary and the many large estates in the county, not only those in the Dukeries but those of the lesser aristocracy and gentry as well, it is surprising that Nottinghamshire did not form more quickly. Maybe the delay arose from the need to first select and appoint a new Lord Lieutenant.

Thoroton's *History of Nottinghamshire* records the event with due respect as follows:

1794. This year is marked by the loyalty of the inhabitants of the Town and County ... Four Troops of Yeomanry and Cavalry were raised out of the most respectable of the inhabitants similar to what was done in other places: their clothing scarlet and buff: their commander Anthony Hardolph Eyre Esq., of Grove near Retford.

Anthony Hardolph Eyre (1757 -1836) was thirty-seven when he took command, the eldest son and heir of one Anthony Eyre (1727–1788). The Eyres were a Derbyshire family which came to prominence in the Middle Ages. This branch of the family had migrated via south Yorkshire to Rampton in Nottinghamshire just outside Retford, where they owned the Manor and significant lands. They also held land in Yorkshire. In 1755 Anthony Eyre married Judith Laetitia Bury, the great niece of Sir Hardolph Wasteney of the neighbouring village of Headon. Through that marriage Anthony Eyre acquired further lands in Lincolnshire and Nottinghamshire. In the 1760s, he expanded his holdings further by purchasing the manor of Grove also near Rampton, together with an estate at Grove, Little Gringley and Ordsall. He sold his Yorkshire estates in the 1860s thus consolidating his land holdings in North Nottinghamshire. By that time he must have been quite a significant land owner.

Anthony Hardolph Eyre was educated at Harrow School and was commissioned into the 1st Foot Guards. He rose to the rank of Lieutenant-Colonel, but had left the Army by the time the Regiment was raised. In 1783 he married Francisca Alicia, third daughter of Richard Wilbraham Bootle Esq. of Latham Hall, Lancashire. They had four children, a son and three daughters, Mary Laetitia, Frances Julia and Henrietta.

Eyre would have been selected for command by the 3rd Duke of Newcastle, as Lord Lieutenant. The involvement of the Duke would also explain why some of the relevant meetings took place in Retford, the least central of all the possible locations. Retford was the Duke's local town; he had relinquished his seat as its MP only weeks before and almost certainly the Retford Troop contained many from the Duke's Clumber Estate and was a collaborative effort between the two men, and possibly others as well. Anthony Hardolph Eyre was obviously a good choice because, from the letters that survive and his speech on the occasion of the presentation of the Regiment's standards quoted later, he comes across as a man of sound judgement.

It is very likely that the 3rd Duke managed the whole process by which the Regiment was raised. He had been Colonel of the 17th Light Dragoons and

Colonel of the 75th of Foot and, as a former major general, was probably the senior and most experienced retired officer in the County. In addition, he came from a family used to being in control of events and would have used the committee chaired by Frederick Montague as the conduit through which he operated. It is almost certain that he would have played a leading hand in the choice of uniform and many other details as well. In summary it is likely that it was the 3rd Duke of Newcastle who was the true architect and founder of the Regiment.

Of those who were named as the first officers of the Regiment the background of only three is known. The first is Cornet Thomas Woollaston White of Wallingwells who went on to raise a volunteer infantry sub-unit which he called the Sherwood Rangers and therefore of whom much more later.

The second is Captain John Denison of Ossington. He was an MP, although it is not known for which constituency, and had originally been called John Wilkinson. He was the nephew and beneficiary of Robert Denison who had in turn been the beneficiary of his brother William Denison, a prosperous merchant from Leeds. John Wilkinson changed his name to Denison on inheriting. In addition to the family home at Ossington Hall, and other lands in Nottinghamshire, the Denison estates included lands in Lincolnshire, County Durham and Yorkshire, as well as businesses in Leeds. John Denison was the first of several members of the family to serve in the Regiment over the next century and a half and, as shall be seen, the father of a most distinguished son, also destined to be a member of the Regiment.

The third of the original three is Captain Ichabod Wright (1767–1862) whose portrait hangs in the Sherwood Rangers' museum. He was one of the proprietors of Wrights Bank in Nottingham founded by his father and which was highly regarded.

An account of his life, recently posted on the internet states:

When the South Nottinghamshire Yeomanry was formed in 1794 he was appointed Captain-commandant of the four troops, and in 1808 he succeeded Colonel Elliott in the command of the Nottingham Volunteers, a force organized in 1798, when threats of a French invasion were rile (sic). His interest in both services continued long after his official connection ceased, and when the present rifle corps, the 'Robin Hoods', was formed, he presented the Mapperley Cup as a prize for the best marksman.

The statement that Captain Wright was commandant of all four Troops is, of course, mistaken, nor was it the South Nottinghamshire Yeomanry which was raised at that time, but it is reasonable to speculate that, since the biggest donor in the south was Baron Middleton, Ichabod Wright would have been approved by him and therefore the Baron may well have banked at Wrights Bank, if so that would have been a sensible arrangement all round.

It is interesting to note that no one from the Dukeries families gained commissions in the Regiment at this stage. There were good reasons why this was so. The 11th Duke of Norfolk was aged forty-eight in 1794 and was therefore too old; and he had no children. The 3rd Duke of Portland was aged fifty-six and heavily involved in Government, and in 1794 his son and heir was only eight years old. Charles Pierrepont was aged fifty-seven. However, he did have four sons, but in 1794 they were aged nineteen, eighteen, sixteen and twelve, and so were still too young. Finally the 3rd Duke of Newcastle's son was only nine years old.

A yeomanry Troop recruited from a much wider radius than did a company of militia. The reasons were logical. Firstly, to find sixty to eighty people, the establishment of a Troop, wealthy enough to own their own horses, it would be necessary to recruit over a much wider area. Secondly, because the individual travelled to the place of assembly on horseback he could come from farther a-field and so it was possible to recruit from a wider area. Typically a troop would be recruited from within a radius of about ten miles or so of its place of assembly because, as for a day's hunting, that was about the limit that a man could ride somewhere, make some significant use of his horse when he got there and then ride home, all in the course of a single day.

There is no doubt that all the estates of the Dukeries, along with others, produced volunteers for the various Troops. Welbeck and Worksop Manor most likely contributed to the Mansfield Troop, particularly as the Welbeck estate reached out to Mansfield. Although nothing specific is known of William Boothby, given that the 3rd Duke of Portland was one of the principal subscribers it is reasonable that William Boothby was his nominee. Clumber and Thoresby, and maybe also Osberton would have contributed to the Retford Troop. Holme Pierrepont contributed to the Newark Troop or possibly the Nottingham Troop and the Newark Troop would have embraced Southwell.

It is significant that the Troops, although recruited mainly from countrymen, were, at this stage, identified with the principal towns by their titles and it is suggested that this was likely to be because it was in the towns that any republican uprising was most likely to start. As will be seen, this would change in later years as their agendas separated as a result of which the towns distanced themselves from the aristocracy and the lesser landowners.

In so far as the French troops in Antwerp were ever perceived to be a credible threat to Great Britain that perception must have receded quickly, as the French force itself soon turned its attention to fighting the Dutch.

If the correct explanation for the declaration of War was that it was a purely opportunistic gesture to trigger a revolution in England, it turned out to be a big mistake. The French would have been far better advised to let the sleeping British military and naval dogs lie in their existing un-mobilized and unprepared state. Instead, as has been seen, the declaration triggered a vigorous preparation and mobilization. Though it soon became clear that there was no credible threat of invasion, this did not bring about a slackening in the rate at which new volunteer units were being formed or the high profile which those units once formed were given. This confirms, as was implicit from the wording of both the legislation and the resolutions, that at this time the most feared threat was that of rebellion at home, from amongst the significant minority who were disaffected republicans.

There is little sign in these early years of any significant input from the military chain of command, or of any professional support from the standing Army, nor even any annual period of continuous training. The absence of these elements indicates that the intended use of these regiments could be met by a lower level of competence than that of a fully trained cavalry regiment. The activities in which the Regiment took part were not full-scale manoeuvres but seemed to mainly consist of high profile parades which often appeared to be prompted in order to provide an expression for the patriotic fervour generated in the County by the fear of invasion and rebellion. These parades also served as a warning to any group of republicans with a mind to try to de-stabilize the monarchy, rather than providing an opportunity to acquire and perfect basic

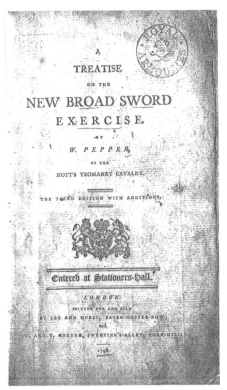

The cover and one page of a comprehensive training manual produced by the Regiment.

military skills. Troops also sometimes provided escorts to some passing dignitary in need of protection.

However there was obviously some attempt made to train, but often these training exercises seemed to be stage-managed affairs carried out very much in the public eye which enabled the Regiment to be publicly complimented in the local press on the progress they had made. On one occasion they were on the Forest in Nottingham going 'through a variety of evolutions to the satisfaction of a vast assemblage of people'. The Forest is an open area on the edge of Nottingham, at that time beside the Mansfield road. On another occasion, Captain Ichabod Wright's Troop 'began the evolutions of the day by going through the Hungarian broad-sword exercise and was followed by the other two Troops of Cavalry, which they performed in such a manner as must ever reflect upon them the highest praise'. Whenever Troops gathered for field days, the day's training always seemed to end with a party. Plainly the Regiment started as it has indeed gone on.

Meanwhile, whilst the Regiment was being formed, the entire country was keenly aware that the pace of the trials and executions in Paris continued unabated and had even extended to include the very architect of the Revolution, Robespierre himself. The feeling of insecurity that this knowledge brought with it was not improved by the knowledge that the standing Army was in a poor state, so that, although the Government was keen to assist the Dutch in their struggle with the French, only 5,000 troops could be spared to send to Holland.

On 17 September 1794, the Nottingham Troop held its first parade in uniform, inspiring the Right Honourable Lord Chief Justice Baron Macdonald in his charge to the Grand Jury at Nottingham to describe the event as

an orderly and dignified preparation by the Yeomanry, or in other words the pith and substance of this country, to resist external force or resist lawless outrage by legal internal force, we see, and look with gratitude and confidence.

Also in September 1794, to demonstrate that the Regiment was fostering embryonic Poet Laureates long before Keith Douglas joined, the following was published which at least has the merit of reflecting the mood of the times:

" If the French shall invade us we'll give them a dose
" We'll fight 'em brave Boys, and we'll fight 'em close
" As Howe fights, and Warren, and Wolfe fought before
" We have beat them so oft, we may beat them once more.
      Chorus—" Long, very long, may Nottingham be
            " Proud of her valiant Yeomanry."

" We fight for our country, we fight for our lives,
" We fight for our sweethearts, we fight for our wives,
" We fight for our children, we fight for good cheer
" And Britons united have little to fear,
      Chorus—" Long may Newark and Southwell be
            " Proud of their glorious Yeomanry."

" In your praise my brave Boys I could carol a year
" You've a gallant good Chief and a Justice in Eyre
" Of Boothby and Wright I with pleasure could sing,
" All firm in support of our laws and our King
      Chorus—" Long may Retford and Babworth be
            " Proud of their noble Yeomanry."

" The Proteus of France may soon change its shape
" And the Tiger, exhausted, return to the Ape,
" While the Banner of Britain which now is unfurled,
" At once may give peace and give laws to the world.
      Chorus—" Long may Mansfield and Worksop be
            " Proud of their gallant Yeomanry."

This parade was followed by field days on the Forest on 26 September and at Thurland Hall on 3 October. Thurland Hall, which stood on the corner of Thurland Street and Pelham Street in Nottingham, had been erected by the Pierrepont family 150 years earlier as the main residence of the owner from time to time of Nottingham Castle and the surrounding estate since the castle was, then, uninhabitable. In case an explanation of why it was chosen as a suitable venue for a field day is required, it had a banqueting hall but, more to the point, was in such a prominent location in Nottingham, that it ensured that those who might contemplate plotting trouble were able to take note of the show of force and think again.

1795 brought military reverses in Europe. First the British force that had gone to the aid of Holland was defeated and withdrawn. This left a small force of French loyal to the British who had seized the vital Mediterranean naval port of Toulon, where a British fleet lay as our only toehold on mainland Europe. Even this fell when the French Army, on the advice of a young engineer lieutenant from Corsica by the name of Napoleon Bonaparte, took the town by securing a fort covering the entrance to the harbour from the sea thus forcing the Royal Navy to withdraw.

The Nottingham Troop is again mentioned by Thoroton as having attended Divine Service at St Mary's Church in Nottingham on 25 February 1795, the start of a longstanding tradition, particularly for the South Notts Hussars.

Chapter 4

The Corn, Bread and Food Riots 1795

Masked by the French Revolution, there were great changes afoot in Britain. Partly inspired by the widely-read Thomas Paine's *Rights of Man*, society increasingly realized the unfairness of the status quo. People saw the massive power and wealth based on agriculture and land ownership vested in a tiny minority of immensely powerful men, who had reigned hitherto without equal. They began to challenge the assumptions on which this order of things was based. The greatest dissent was felt in the towns and cities where wealth and power was developing based on trade and commerce, which was relatively much more evenly spread through a growing meritocracy. The first manifestation of these pressures, which ran in parallel to the Napoleonic threat, so far as the Regiment was concerned, were the hunger-inspired corn, bread and food riots, precisely the same pretext that had triggered the French Revolution.

In April 1795, eight months after the Regiment had been raised, riots broke out in Nottingham, among many other places. Triggered by a different, but related threat. The cause was the rise in the price of wheat to £5 a quarter ton, equivalent to £250 today. Bad weather, which again may have been due to volcanic activity in Iceland, had caused a shortage. Unsurprisingly, the ability of landowners to apply spontaneously this sort of price rise for vital basic staple foods became a major political matter. Indeed it was destined to define the political struggle between the landowners and the emerging meritocracy for the next half century. In the early stages it led not only to rioting, but more subtlety, and indicative of the country's increasing commercial awareness, to the import of cheap wheat from abroad.

This caused the landowners to respond to the unrest by ensuring that the yeomanries, manned, of course, by those loyal to them, were called out to contain the riots, and to respond to the import of cheap wheat by arranging for legislation to be carried swiftly through Parliament to prevent such imports, which again triggered further riots. The legislation, known generically as the Corn Laws, was a huge bone of contention and became a symbol of this struggle, and was eventually repealed in the 1850s amongst much rejoicing.

The Nottingham Journal contained the following account of the riots of 1795, which must have had everyone watching its every twist and turn for fear that this could be the tinderbox for a full blown revolution if not managed well:

Saturday evening last, serious consequences were apprehended in this Town by the tumultuous assemblages of the populace, who threatened to commit some depredations on the property of butchers, &c., on account of the present high price of all kinds of the necessaries of life. The Magistrates in consequence immediately assembled in council to deliberate about the most proper measures to be taken to secure the tranquillity of the Town; when it was thought expedient to have the Proclamation in the Riot Act read, but finding all their attempts ineffectual after this, the Chief Magistrate requested the attendance of the Troop of Gentlemen Yeomanry Cavalry (the Nottingham Troop); and never did any set of Regular Troops obey the call with more cheerfulness and alacrity. For no sooner was the signal given for their attendance, than nearly the whole troop assembled in the Market Place, fully accoutred, and ready for immediate service; which reflects the highest honour upon them for their attention to the security of the Town.

By their diligent exertions for nearly four hours, added to that of a troop of the Heavy Dragoons, and the securing of thirteen of the ringleaders, the tranquillity of the Town was restored without any material injury to any person or property.

Monday last the above men (and also another person, who was committed to the House of Correction on Saturday morning, for exciting the populace to riot and was committed to Gaol for want of securities to keep the peace) were examined; when William Stubbs, charged with throwing a stone at one of the Gentlemen of the Troop, was committed to take his trial at the next Assizes.

A story is told that on this occasion a part of the mob took refuge behind some iron railings in the town from where they were able to stone the troopers. Cornet Hadden, much to the surprise of the mob, jumped his horse most gallantly over the railings landing amongst the mob which quickly then dispersed.

In April 1796, the Nottingham Troop had to turn out again to disperse a mob in the town; once again the reason for the disturbance was the price of wheat. The cause was a rumour that a baker called Smart was going to Newark to 'enhance the price of corn', it being alleged that he already had a large stock of wheat in his possession.

The following account of the disturbance appeared in the Nottingham Journal:

On Monday last a disposition to riot manifested itself amongst the populace of this Town, by assembling about the house of Gervas Smart, baker, demolishing his windows, and threatening to pull down his house, etc.

After the Magistrates had used all their influence, without effect, in order to make them disperse, the Nottingham Troop of Yeomanry Cavalry and a party of the 12th Light Dragoons were ordered out to the assistance, who behaved with the greatest coolness notwithstanding the repeated insults they received from the populace; but at length after parading the streets for several hours, and not being able to compel them by fair means to retire to their several habitations, they received orders from the Chief Magistrate to fire upon them, and to the humanity of the above Troops may be attributed the saving of a number of lives, being willing to show them all the levity in their power. We hear that only one shot was fired upon them, which wounded a boy in the foot.

This action caused the mob to disperse so that by 10.00pm order had been fully restored. Between them the Dragoons and the Yeomanry arrested seven people who were then confined in the gaol.

Chapter 5

Presentation of Standards 1795

As has been mentioned the 3rd Duke of Newcastle died unexpectedly on 25 May 1795, to be succeeded as Lord Lieutenant by the 3rd Duke of Portland, an already very busy man, leaving Anthony Hardolph Eyre in sole command of the Regiment.

> The King's birthday, 4 June 1795, was celebrated as a field day by the Nottingham Troop with the Nottingham Journal reporting that 'upwards of 5,000 spectators ... came to witness the hilarity of the day: at this instant the Yeomanry returned from the Forest to the Market Place and fired three volleys on horse-back, in a manner which great praise is justly due; ...

Twelve days later the whole Regiment met together for the first time for drill and manoeuvres at Thoresby Park, the home of the yet-to-be-ennobled Charles Pierrepont MP, who was stepping into the void, in relation to the Regiment, left by the untimely death of the 3rd Duke of Newcastle, a responsibility which Charles Pierrepont (notwithstanding the fact that, as has been mentioned, he had served with distinction in the Navy rather than the Army) and his family took on and discharged with great commitment over the years ahead.

The Nottingham Journal reported that 'The day being remarkably fine, it drew there an immense concourse of spectators from all the towns and villages in the neighbourhood. They [the Regiment] went through their various manoeuvres for upwards of four hours with such military exactness that gave universal satisfaction to every beholder'. It will come as no surprise that 'after the toil of the day was over, they were invited to a most sumptuous entertainment in the Hall'.

On 14 July 1795, Standards were presented at a magnificent ceremony in Nottingham. Thoroton, in his county history, relates how on

> Tuesday last the respective troops of Nottinghamshire Yeomanry Cavalry (comprising the Nottingham, Newark, Retford and Mansfield troops) met

together to receive their colours. The day proving exceeding fine it prompted an innumerable concourse of spectators to view the novelty of the scene. About ten o'clock the troops took their ground in Sneinton Field, from whence they rode in regular military procession to the Market Place and, forming a square in front of the Exchange Hall, the windows of which, being filled by ladies of the first rank and fashion, the sight became truly enchanting. Everyone seemed pleased, and doubtless admired the patriotic spirit of their countrymen...

Charles Pierrepont, presented the Royal Standard in the name of Henrietta Lumley-Savile, the daughter of the 5th Baron Middleton of Wollaton and wife of Richard Lumley-Savile of Rufford Abbey and therefore destined to become the Countess of Scarbrough. Thomas Webb Edge, Esq., presented the Provincial Standard. This presentation was in the name of Lady Warren, the married daughter of a General 'of distinguished reputation'. It is likely that not only were the Standards presented by the ladies in question but that they were respectively responsible for their making. The Royal Standard is crimson with the Royal Arms embroidered on each side. The initials NYC (Nottinghamshire Yeomanry Cavalry) are also included. The Provincial Standard is of buff silk with, on the one side, the Royal Cipher (GR) embroidered in the centre, ornamated with roses, wheat ear and olive branches intertwined. The Arms of the City of Nottingham are included as well as the Latin mottoes 'Freedom under a just king' and 'For wives children and homes'. On the reverse is an oak tree with the Latin motto 'Both honour and protection' and the words Nottinghamshire Yeomanry Cavalry. The crimson relates to the Regiment's tunics and the buff to the facings.

Lieutenant Colonel Eyre's speech, as part of the ceremony of presenting the Standards to his Regiment is worth recording because it gives a clear idea of the mood of the times and the perceived threat they were facing:

Gentleman,- In the names of Mrs Lumley-Savile and of Lady Warren I have the honour to present you the Standards of the Regiment, which it is your duty to defend with your lives. I flatter myself that few exhortations will be necessary to induce you to fill this duty, when you consider the cause in which they are set up – the cause of your King, of your Constitution, of your Religion, and everything that is dear to man or sacred to God. A neighbouring nation having torn asunder all the bounds of civil society, having trodden under foot all laws human and divine, has dared in the hour of her insolence to threaten this country with invasion, and relying for assistance on the traitorous promises of some disaffected individuals within this realm, has ventured to hope that she might plant her destructive principles in this soil; but I trust that the universal loyalty and attachment to the Constitution which have been manifested through the kingdom, will convince her of the folly of her expectations and that we shall secure to ourselves peace and tranquillity by being prepared for war.

It must give the most heartfelt satisfaction to every good citizen to see the number of Volunteers who, at this alarming crisis, have stood forward in support of our country, and who have shown themselves worthy of the blessings we enjoy under our present form of government, by being ready to sacrifice everything in its defence. With spirits such as yours my comrades I will be bold to say we shall overcome all our foes, foreign and domestic; we shall support our laws, maintain our liberties, and transmit to our posterity that excellent Constitution which has been established by our ancestors after many hardy contests, and which has long been the envy and admiration of the world. For this cause, gentlemen, our Standards are now erected; of this cause who does not feel it is his duty to die in defence? And when you consider the fair hands from which you have received them and the smiles of beauty yield us their patronage, I am convinced you will all feel what is your duty, your delight.

Both these standards still exist and are in the possession of the Regiment.
After the consecration, the account continues:

The Regiment then marched to Sherwood Forest, (after some training) the Corps dined in Thurland Hall, where were given suitable and loyal toasts. The happy day ended with a ball and with fireworks displayed in the Market Place.

By the end of 1795 Bonaparte had been placed in command of the French Army and in 1796 he conducted a brilliant campaign, invading northern Italy, then occupied by the Austrians and there defeated them. He then turned his attention towards Egypt via which he planned to seize the whole of the Orient.

On 1 September 1796, Lieutenant Middlemore and Cornet Hadden of the Nottingham Troop were presented with the freedom of Nottingham, presumably in recognition of their actions on the two occasions on which the Troop had been called out the previous year. It was in 1796, also, that Charles Pierrepont was ennobled and thereafter was addressed as Lord Newark.

In 1797, on 6 June, the whole Yeomanry paraded together at Mansfield for inspection, the inspecting officer being Major Gray, during which, according to the Nottingham Journal 'they went through their various evolutions with the exactness of veteran troops'. As you will have guessed this occasion ended with a dinner for all ranks at the Moot Hall.

Chapter 6

Raising the Sherwood Rangers 1798

By now the Government's efforts to re-arm were beginning to bear fruit so that the Royal Navy was strong enough to hold its own but, even so, the process was a race against time because, on the Continent, Spain was in alliance with France, the Netherlands were now little more than a French satellite, and that part of the Low Countries which would become Belgium had simply been annexed. Austria and even Venice were now also subservient. Britain was almost alone.

At home things were little better. The republican movement was proving so problematic that in order to suppress it the Government had been driven to suspend the right of Habeas Corpus, that most fundamental of human rights. How little do things change? The Government had responded in other ways as well. For example, in the spring of 1797 an act called the Provisional Cavalry Act had been passed to address the fact that not all counties had raised yeomanries and some had not raised enough. In outline its terms imposed an obligation on those who owned horses but who had not joined the yeomanry to be formed into regiments and provide their own horses and equipment, not on the restricted terms of service of the yeomanry, but on the same unrestricted basis as the militia. The message was plain and so although few units were raised under the Act it triggered the 'voluntary' raising of further troops of yeomanry – a clever piece of legislation.

With such a tense national internal security situation, also reflected locally, it is hardly surprising that 1798 saw several developments in Nottinghamshire. A Troop of Yeomanry was raised by the Pierrepont family at Holme Pierrepont, commanded by the eldest son Evelyn Henry Frederick Pierrepont, or it may be that sponsored rather than commanded would be more accurate, because there is no record of his commission being gazetted and the Captain of the Troop was one Thomas Bettison. Evelyn Pierrepont was now aged twenty-three. In 1796 he had in addition become MP for Nottinghamshire. Other Troops were also raised at Bingham and Bunny. Whilst the Bingham Troop was incorporated into the Nottinghamshire Yeomanry, the Bunny Troop had been formed on terms of service that limited it to service in its own District and was independent.

An independent troop was also raised at Worksop, which was the cavalry portion of the Worksop Volunteer Association, raised to act in the immediate neighbourhood of Worksop. Its uniform was a scarlet tunic with blue facings. The Troop was commanded by Captain John Froggatt with Isaac Wilson as his Lieutenant and John Roe as his Cornet. It is not known who inspired the formation, but the most likely family would have been the Foljambes at Osberton.

A number of infantry units were also raised around this time in response to the same period of increasing tension, including, crucially for the Regiment, one raised by Thomas Woollaston White, who had originally joined the Mansfield Troop when it was first raised. The unit was based on his estate at Wallingwells and he called it the Sherwood Rangers.

The White family originate from Suffolk, their earliest traceable antecedent being one Bartholomew le Wite who lived in Stoke by Nayland in the thirteenth century. The first record of them living in Nottinghamshire is that of Johannes White of Colyngham in the fifteenth century. By the sixteenth century they had acquired the Manor and lands of Tuxford, which the family has regarded as its spiritual home ever since and where several generations are buried. Their prominence as a family can be gauged by the fact that they were sufficiently well thought of by Mary I for her to grant them three manors, two in Somerset and the third at Cotgrave in Nottinghamshire, and by the fact that they are kinsmen of a number of prominent dynasties, including the Cecils of Burghley, the Dukes of Kingston, and the Viscounts Galway.

They were at times involved politically, two members of the family having represented East Retford. Of the two, the more distinguished was John White, born in 1699 who succeeded to the family estates in 1732 and was elected the member for East Retford the next year. He was a Whig and a supporter of Henry Pelham and the 1st Duke of Newcastle. According to Horace Walpole, he was a member of their inner circle and exerted considerable influence. Ironically, given his son's transparent loyalty to the Crown, it was said that he had distinctly republican leanings.

One of the family's most important and self confessed skills was that of marrying heiresses; indeed they had acquired the Wallingwells estate by marriage when Thomas White's great grandfather, also Thomas White, had married Bridget Taylor, the heiress of the estate in 1698. His heir, Taylor White, married Sarah Woollaston, the co-heiress of Sir Isaac Woollaston of Lowesby Hall, Leicestershire.

The Woollastons were also adept at marrying heiresses. Whether or not it is the done thing to count up the number of heiresses a family has married, the Woollastons had, and had reached an impressive total of twenty-six. As a result Thomas, who was born in 1767 and christened Thomas Woollaston White to acknowledge the family connection of his mother, was the heir to a very large and widespread fortune. He inherited, aged twenty-eight, on the death of his father in 1795 approximately 11,000 acres, about 4,500 of which were in Nottinghamshire. He also inherited lead mines at Pately Bridge in the West Riding of Yorkshire which for a period yielded £1,000 per week (in excess of £100,000 per week in today's values). In addition there were other significant streams of income.

Initially Thomas served in the 4th Light Dragoons, but resigned his commission on the death of his father and simultaneously joined the Nottinghamshire Yeomanry. Soldiering was not his only interest. He was fond of hunting and other athletic sports and was a keen falconer. In pursuit of this interest he kept a large mews and extensive stables and kennels and employed a German falconer by name of Bekkar. Meets of his hawks were arranged every day of the week. His house was renowned for its open hospitality.

Wallingwells Hall is near Carlton-in-Lindrick and lies on the Nottinghamshire/Yorkshire border. The estate itself was only about 500 acres and was embraced by the villages of Woodsetts, Gildingwells and Letwell, all of which are in Yorkshire, due east of Carlton-in-Lindrick. Presumably they made Wallingwells their seat despite the relatively small amount of land surrounding it because of its historic origins as a nunnery built in the reign of King Stephen, and the fact that it was a more substantial house than any others they owned.

What was quite exceptional about Thomas White's decision to raise an infantry unit, compared with all others described in this account is that, whereas all others who raised units, whether yeomanry, or militia, did so in accordance with the legislation which only involved relatively affordable levels of personal cost, he raised, clothed and armed the Sherwood Rangers at his own personal expense. So exceptional was this act that it came to the attention of George III who said the undertaking was too heavy an expense for a private gentleman to bear and offered to share half of the cost with him from his privy purse. Thomas White declined His Majesty's offer, saying he considered it the duty of every loyal gentleman to assist to the utmost of his means at such a crisis. One can only speculate at what he sought to achieve by such a gesture. It might have been as altruistic as it seems. However, he might have felt it removed any lingering doubt, in high places, concerning the family's loyalty to the Crown, given his forebear's republican leanings.

'Rangers' is a term which originates as the name given to those appointed to police the use of forests, often Royal Forests, of which Sherwood Forest had been one. As their name implies, their role was a mobile one to deter or detain those who hunted deer without permission, took timber without a licence or grazed animals without the right to do so. In the military context it had become the name traditionally given to volunteer infantry sub-units used by the British for reconnaissance and skirmishing, particularly in forests and in close country, to which neither cavalry nor line infantry battalions were suited. They used a traditional curled hunting horn with which to communicate rather than unwieldy drums and were used extensively by the British in America during the War of Independence and, particularly, during the French and Indian War of 1755–1763 where they were raised from amongst the indigenous population and called Provincial Rangers. The line infantry regiments, which had a much more structured and limited role, did not trust these native sub-units and, encouraged by General James Wolfe, gradually replaced them with specially-formed light companies of their own, attached on a scale of one for each infantry battalion, an idea which had first been tried in the 1740s. At this stage their tunics were still scarlet.

Following the end of the American war the tactical doctrine involving the use of light companies continued to develop and eventually, by 1802, had realized its full potential under the guiding hand of General Sir John Moore and was then used to powerful effect as dismounted manoeuvre troops during Wellington's Peninsular campaign which commenced in 1808 and which ended with the initial defeat of Napoleon in 1814.

In its fully developed form, the doctrine called not only for light companies to be enlarged into battalions, but that they also be equipped with the newly-developed rifled barrelled weapon, which was much more accurate than the smooth-bored flintlock. These new regiments were of two different genres. The first called themselves the Light Infantry and adopted the horn and lanyard as their insignia and dark green for concealment as their uniform. The second chose to emphasize their exceptional skill as marksmen and consisted of two regiments, the Kings Royal Rifle Corps, the 60th Rifles, and the 95th of Foot which became known as the Rifle Brigade and adopted the same dark-green uniform but different regimental insignia, particularly their distinctive black buttons. For communication both exchanged the curled horn for the bugle. Together these regiments became famous, initially under General Moore's command, as the Light Division.

It is not unreasonable to deduce that Thomas White took his soldiering quite seriously and had reached the conclusion that a light infantry unit of the type that had been so successful in the War of Independence was better suited to the forests of north Nottinghamshire than cavalry and, having become an enthusiast for the concept, decided to form such a unit himself. He may well have continued to serve in the Nottinghamshire Yeomanry as well, since in 1797 he had a Nottinghamshire Yeomanry officer's uniform made which still belongs to the White family.

Thomas White had a barrack built in the park to Wallingwells which provided a base for his unit to use for training. The uniform he adopted for the Sherwood Rangers, and used at least until 1802, consisted of a jacket of superfine green cloth with a stand-up collar. It had cuffs of the same cloth, edged and feathered with white cassimere (a plain or twilled woollen cloth used for men's clothing), and, for the officers, trimmed with silver vellum lace holes. It had wings (epaulettes) that incorporated a small horn and lanyard within an oval. The White family still have Thomas Woollaston White's uniform dated as having been made in 1798.

Every regiment that has adopted the insignia of the horn and lanyard as its cap badge has designed a unique variation of the theme but the horn and lanyard insignia as worn by the Sherwood Rangers is identical to that worn by the Oxfordshire and Buckinghamshire Light Infantry. Since neither of its antecedent regiments, 43rd and 54th, became light infantry until 1803 this suggests that the Sherwood Rangers are entitled to claim first use of the design.

Meanwhile on the big stage: in 1798 Napoleon sailed to invade Egypt with Nelson in hot pursuit, determined to prevent him from realizing his ambitions in the Orient. Dancing over the waves in the company of Nelson's mighty men o' war was the diminutive *Kingfisher*, too small to take part in the forthcoming battle itself, but of significance because she was commanded by Charles Arthur Herbert Pierrepont (1778–1860), the third son of Lord Newark. He had joined the Navy and had just that year been appointed post captain at the early age of twenty. Coincidentally, it was an achievement identical to that which his father had enjoyed some forty years earlier. He was destined to command the Holmpierrepont Troop.

On reaching Egypt, Nelson found the French Fleet at anchor, line ahead in Alexandria harbour hard against, therefore believing its flank to be protected by the shore and with all its guns facing seaward. On 1 August 1798 Nelson attacked and, to the disbelief of the French, sailed his fleet between them and the shore where the French thought there was insufficient water, free to fire broadside after broadside on them with little response. Nelson's victory was total, leaving the British in control of the Mediterranean and Napoleon and his army stranded in Egypt.

At home, the Nottingham Troop was so successful that in 1798 it built its own riding school. The engraved headstone for this building is in the possession of the Regiment.

Napoleon managed to escape from Egypt in 1799 and in 1800 attacked and defeated Austria at the Battle of Marengo.

Chapter 7

The Corn, Bread and Food Riots 1800

In 1800, whilst Napoleon was trying to recover from the effects of his defeat at the Battle of the Nile, bread riots, as they had become known, again occurred in Nottingham, for which the Yeomanry had to be called out. The first, which occurred on 19 April, started in the Market Place where a considerable number of people seized a large quantity of butter and potatoes and, after destroying a number of stalls and market carts, attacked the butchers and took away a quantity of their meat. On the militia and the Nottingham Troop being called out, order was restored quickly, and a number of people were arrested.

Very bad weather was experienced that summer, causing the price of grain and other provisions to rise very sharply to levels that few could afford. These price increases were, therefore, matched by an equally sharply rising mood of discontent. Matters came to a head on Sunday 31 August when a spontaneous gathering took place in the town centre and rioters commenced breaking into some bakers' premises and attacking the granaries at the canal wharves. The Volunteer Infantry, three troops of the Blues and the Nottingham, Holme Pierrepont and Bunny Troops of the Regiment, were called out. The London Morning Chronicle subsequently published the following account of what then took place:

Nottingham, Tuesday, 3 o'clock – It is very painful for us to state that the public discontent is still violent, and there is not even a prospect of the riot subsiding. The Yeomanry Cavalry, and the Infantry belonging to this town, have been on duty all this day, and a great part of yesterday; but the poor inhabitants, who are alone discontented, absolutely defy them, and even the women interrupt them with hisses, and other modes of abuse. About 14 persons have been secured and are now in Nottingham Gaol. The women are the principal aggressors and they are permitted to remain at liberty. This morning, about 1 o'clock, a large party attacked a baker's shop, near Sutton's the bookseller; they ransacked the shop of all the flour they could find, which they exhibited to the persons present as containing chalk, alum and other poisonous substances. A Troop of the Blues, however, arrived, and after the greatest exertions succeeded in disbursing the mob. The Blues have been unremittingly on duty three days and two nights, without rest; and they must continue to parade the streets until reinforcements shall arrive. Many of the shops are shut up at midday, and a real panic pervades the minds of the opulent. The rioters at Mansfield have been overawed, and that place is now quiet; but they are so numerous in Nottingham, and the want of bread so general amongst the poor, that unless something is done to relieve them, the consternation must continue.

This morning about 12 o'clock, the flag was hoisted at the top of St Mary's steeple and a messenger sent to Sir Thomas Parkyns (father of Lord Ratcliffe) to request he would immediately dispatch his troop of horse from Bunny, a village in the neighbourhood; they arrived about an hour ago, and are now parading the town.

A large party made an attack upon a mill in the environs; they loaded a wagon with wheat, and were making off with it when a party of horse arrived, rescued the wheat and, for greater safety brought it to the Town.

The vengeance of the rioters is directed entirely against the bakers, millers and farmers; they attribute the present scarcity to forestallers, monopolists, and regrafters. Business is entirely at a stand. All the weavers have joined the crowd.

Five o'clock – This moment between twenty and thirty of the rioters have been lodged in the County prison; they arrived in an open wagon, under an escort from the village of Arnold, about four miles from Nottingham and all belonged to the extensive manufactory of Messrs Davidson and Hawkesely. Many of them are fine young fellows, and appear under twenty years of age. The women would gladly have affected a rescue, but were overpowered by the soldiery. There was a serious conflict before these prisoners could be secured, and the Bunny Yeomanry were obliged to fire on them; one is dreadfully wounded with a ball in the neck, and another had his arm shattered.

The riots continued for three days and, ironically, since bad weather had been the cause in the first place, the final dispersal was in great measure due to one of the most terrific storms of thunder, lightning and hail that had ever been witnessed in the locality.

The following notice was inserted in the Nottingham Journal of 6th September by order of the Magistrates of the Town:

TOWN OF NOTTINGHAM

The disposition to riot and tumult happily appearing to be now subsiding, the Magistrates have great pleasure in returning their thanks to the officers and privates of His Majesty's Regiment of the Royal Horse Guards Blue, and to the different Corps of Yeomanry and Infantry co-operating with them, for their temperate and active conduct in support of the Civil Power.

The Magistrates have been anxious to take every step in their power for the immediate restoration of public tranquillity, and in that view they thought it their duty to announce to the public (by hand bill) that they had reason to believe a quantity of corn would be brought into the Market and sold on Sunday next at £4 per quarter. They now feel bound in justice to the liberty of the Nottingham, Bunny, Holme Pierrepont, and Ratcliffe Troops of Yeomanry Cavalry, and the Nottingham Volunteer Infantry Companies (commanded by Lieut. Colonel Smith, and under him Major Hooley, Robert Padley and Samuel Deverill, Esquires), to state to the public that the above communication was made with the authority of the different Farmers, and Gentlemen composing these Corps, who came forward to endeavour by their own exertions to indulge in the public in general to take effectual measures for bringing down the price of grain.

By Order
GEO. COLDHAM
Town Clerk

Probably there was trouble in the other towns in the county as well, but records are only complete for the county's capital.

All yeomanries were rewarded for their services in relation to quelling the prevailing civil disorder by being exempted from the hair-powder tax. This magnanimous gesture, no doubt, caused a veritable stampede of volunteers from amongst the farmhands of England. Serving yeomen were also exempted from the horse tax, a much more useful measure.

On 14 October 1800, the Retford Troop was ordered to parade on 17 October, in front of Captain Lumley-Savile's home, Rufford Abbey, to escort HRH the Prince of Wales, the future George IV, some miles on his journey. The Troop was ordered 'to be in scarlet jacquets, hair well powdered, cloaks well roll'd and their horses, accoutrements and arms in the best possible order'. 'Hair well powdered.' Is that irony? Surely not.

In March 1801 Pitt resigned, making way for an ill-fated Government of National Union, but which was in fact of a mainly Tory persuasion, under Addington. The same year the British, under Abercromby and Nelson, invaded Egypt and defeated the French army trapped there, thus ending for ever

(Left) Lt Col E. O. (Flash) Kellett DSO MP, commanded the Sherwood Rangers Yeomanry 1940–43. Killed in action on 22 March 1943.

(Right) Lt Col J. D. (Donny) Player, commanded the Sherwood Rangers Yeomanry in 1943. Killed in action on 24 April 1943.

Lt Col Cmdt Anthony Hardolph Eyre. Raised and commanded the Nottinghamshire Yeomanry 1794–1802.

Captain Sir Thomas Woollaston White, 1st Bt. Raised an infantry unit named the Sherwood Rangers in 1798 having joined the Nottinghamshire Yeomanry in 1794. (By kind permission of Sir Nicholas Woollaston White)

Cornet Alexander Haddon and the Nottinghamshire Troop confronting rioters in Nottingham during the Bread and Food Riots, 1795. (By kind permission of the South Notts. Hussars)

Captain the 2nd Earl Manvers commanded the Holme Pierrepont Troop of the Nottinghamshire Yeomanry. (By kind permission of the Manvers Trustees)

JEREMIAH BRANDRETH.
Beheaded for High Treason at Derby.

The late Jeremiah Brandreth executed for treason as a result of his role in the Derbyshire Rising, also known as the Brandreth Riots.

Captain Henry Charles Howard, Earl of
Surrey and later 13th Duke of Norfolk who
commanded the Worksop Troop. (By kind
permission of the Duke of Norfolk)

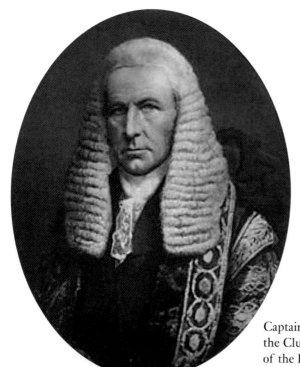

Captain John Evelyn Denison commanded
the Clumber Troop. Later became Speaker
of the House of Commons and 1st Viscount
Ossington.

4th Duke of Newcastle–under–Lyne at the time of the Reform Bill Riots.

Lt Col the 5th Duke of Newcastle commanded the Sherwood Rangers Yeomanry 1852–1864 and was Lord Lieutenant 1851–1864.

Lt Col S. W. Welfitt commanded the Sherwood Rangers Yeomanry 1864–1878.

Captain John Vessey Machin commanded the Clumber Troop. (By kind permission of the Machin Family)

The 5th Duke of Newcastle entertaining HRH the Prince of Wales in some style on the lake at Clumber Park, 1861. (Illustrated London News)

HRH the Prince of Wales alighting at Clumber Park attended by the Clumber Troop, 1861. (Illustrated London News)

HRH the Prince of Wales escorted through Worksop by the Clumber Troop, 1861. (Illustrated London News)

Worksop Manor. (By kind
permission of Bassetlaw Museum)

Clumber Park. (By kind
permission of Bassetlaw Museum)

Wollaton Park. (By kind
permission of Lord Middleton)

Thorseby Park.

Bonaparte's ambitions in the Orient. This was such a great victory over the French that celebrations took place all over the country. In Nottingham the event was commemorated in November by the Nottingham, Holme and Bunny Troops parading in the Nottingham Market Place and marching to St Mary's Church.

This British victory in Egypt created a favourable change of climate as a result of which a peace treaty was signed at Amiens on 28 March 1802.

This is a synopsis of all officers recorded as having served in the Regiment to this point:

Troop	Name	Date Gazetted
Retford Troop	Captain Anthony Hardolph Eyre of Grove.	15/8/94
	Lieutenant Samuel Crawley	15/8/94
	Cornet Robert Ramsden	15/8/94
	Cornet John Charles Giradot	5/4/98
Mansfield Troop	Captain William Boothby of Edwinstowe	15/8/94
	Lieutenant John Sutton of Scofton	15/8/94
	Cornet Thomas Woollaston White of Wallingwells	15/8/94
	Lieutenant John Smith Wright	25/7/99
	Cornet William Wylde	25/7/99
Newark Troop	Captain John Denison of Ossington	15/8/94
	Captain Francis Chaplin of Newark	21/11/94
	Lieutenant Phillip Palmer of East Bridgford	15/8/94
	Cornet Francis John Brough of Newark	15/8/94
Nottingham Troop	Captain Ichabod Wright of Mapperley	15/8/94
	Lieutenant William Richard Middlemore	15/8/94
	Cornet Alexander Hadden	15/8/94
Holmpierrepont Troop	Captain Thomas Bettison	20/6/98
	Lieutenant Christopher Jowitt	20/6/98
	Cornet Tempest Haines	20/6/98
Agent	Cox and Greenwood, Craigs Court	

Chapter 8

Peace in Our Time 1802

The Treaty of Amiens came into force on 1 May 1802. As a result there was a considerable relaxation in tension, to such an extent that many people even became tourists and flocked to France to see the scenes of the Revolution.

One of the consequences of the Treaty was that all yeomanry troops were ordered to disband. Given the civil unrest with which the Regiment and other yeomanries had had to deal, which was only indirectly linked to the French Revolution and Napoleon's early offensives, that decision was surprising. However, the Mansfield, Retford, and Nottingham Troops duly complied. The Newark, Holme Pierrepont and Bunny Troops, on the other hand, remained active, suffering from the selective deafness in relation to orders they do not like which has, from time to time, afflicted the yeomanry throughout its existence. Given their experiences of the previous five years it is easy to understand why they would have felt the order was unwise.

In Nottingham a dinner was held in the Riding School at which Lord Newark presented honorary medals to each member of the Nottingham Town Troop in recognition of their important service. The three officers received gold medals whilst those for the men were of silver. The King's Head and a picture of the Greendale Oak appear on these medals together with the motto '*Foi, Loi, Roi*'. No doubt it was the Greendale Oak's pre-eminence at that time which caused it to be selected by Lord Newark when he designed the medals.

Where the history of Sherwood Forest is concerned, fact and mythology compete with each other like two growths of ivy entwining their way round one of its great oak trees. No more clearly is this illustrated than through the stories attaching to the most famous of those trees of which there were four. The best known is the Major, or Queen's Oak, which is the only one standing to this day, albeit, stayed like a prisoner in chains. It is this oak which is said to have provided Robin Hood with sanctuary when on the run from the Sheriff of Nottingham. The next is the Parliament Oak, the location of which is no longer marked on maps. Its claim to fame is derived from the legend that King John gathered his ministers in session under its mighty boughs, Cabinet meetings outside the capital, therefore, not being the original idea that recently they have been claimed to be. Then there was Robin Hood's Larder Oak, also not marked. The mightiest of them all, however, was the Greendale Oak which stood for over 800 years half a mile south of where Welbeck Abbey was eventually built. To give some idea of its size, in 1724 a coach road had been driven through its massive trunk creating an aperture measuring 10-feet 3-inches high by 6-feet 3-inches wide. Its site, which is still marked on some maps, is now occupied by two young oak trees grown from its acorns.

Thomas White also disbanded the Sherwood Rangers, and was rewarded by George III for his service to the Nation in raising a unit wholly at his own expense, by being created a baronet on 30 November 1802 with remainder to his heirs and those of his brothers, a title which exists to this day. The coat of arms he selected contained the motto 'Loyal until Death', representing the key sentiment of the day and perhaps, as already mentioned, the key motive for such a noble act of service to his Monarch.

If there really ever had been peace it was short lived. With the benefit of hindsight the French had probably only acquiesced in the peace process to enable them to recover their military poise from the reverses inflicted on them by the British in Egypt and to plot their next move. Bonaparte now resolved that the British must be defeated if he was to be able to pursue his imperialistic ambitions in Europe. Accordingly, in May 1803, he gave orders for an invasion of England and, almost immediately, the preparations taking place as a result on the other side of the Channel became clearly visible from the coast of England and were mirrored by urgency here.

This time the threat was a very real one based on a sound plan and fully resourced. Napoleon's plan was to concentrate a flotilla of vessels suitable as troop ships at a single point, Boulogne being the one chosen, from where he believed it would be possible to convey an Army which eventually grew to 200,000 men, across the Channel with all its equipment and logistical support. He knew that there was, however, one vital precondition for success; that his navy had to have command of the Channel for at least the duration of the invasion itself. It was not an easy precondition to secure, because the Royal Navy was by now far superior and in every way more formidable than his own; therefore he would have to either outmanoeuvre it or suppress it before invading.

The British Government was more than aware of the position and immediately ordered the Royal Navy to sea with orders to impose a blockade on the French navy before it could itself get to sea. The Royal Navy complied with a vice-like efficiency which confined the French navy to its various home ports. As an indication of the commitment with which the Navy carried out its orders, it was to be two years before Nelson would set foot on dry land again, two years during which the stalemate created by the blockade held. But the Government knew that all Napoleon would need was a few precious days in control of the Channel and he would be across and ashore, so the margin for error was very small and the Nation felt its fate hanging by a thread. These were tense times.

Amongst a host of other preparations to defend the island, both great and small, Downing Street hastened to write to the Lords Lieutenant on 30 July 1803 with plans for the mobilization of reserves. In Nottinghamshire this was received by the 3rd Duke of Portland who still held that office and was also serving in the Government as Lord President of the Council.

The letter read as follows:

> in obedience to His Majesty's Commands ... and to enable His Majesty more effectually and speedily to exercise His ancient and undoubted Prerogative in requiring the Military Service of His liege Subjects in case of Invasion of the Realm and to direct your Lordship to take immediate measures [to raise volunteers]
>
> the Officer in the command of the District will be directed to furnish your Lordship with a Plan of instructions for Drill and Exercise ...
>
> ... In all places where Volunteer Corps can be formed ... every encouragement should be given for that purpose ... to render it unnecessary to have recourse to the compulsory clauses of the Act.
>
> It can scarcely be necessary for me to point out to your Lordship the difficulty of issuing Arms from His Majesty's stores for the extensive training and exercise required under this Act, without material injury to the other essential branches of the Military Service; I must therefore earnestly recommend to your Lordship to resort to the Zeal and Public Spirit of the Inhabitants of the county under your charge for purpose of procuring a return of the Arms in their possession, in order that, with their consent, they may for a time be applied to the Service of the County, and that your Lordship would take measures for distributing them in the manner that it may appear to you most conducive to the objective in view; twenty-five firelocks being considered sufficient for the purpose of drilling One Hundred Men.

Note the phrase 'in case of invasion of the realm' and the lack of any mention this time of 'riots and tumults'.

Presumably the arms referred to were those issued in response to the previous period of crisis which had never been handed back. However, perhaps unsurprisingly, this did not prove to be a practical idea and in the end the Government had itself to make considerable numbers of additional Arms available. It is interesting that the instinct of Governments to mobilize and commit troops to action without providing them with the necessary equipment is not new.

The arrival of the letter had clearly been anticipated, because on the same day a meeting of the Nottinghamshire Deputy Lieutenants held under the Defence and Security Act took place at the County Hall in Nottingham. The Deputy Lieutenants were performing their traditional role. The meeting ordered that 'precepts be issued to the Chief Constables conformable to the directions of the above Act' and the effect of this resolution was to require Chief Constables to compile a complete list of all men in the County between the ages of seventeen and fifty-five. This was a legal requirement and the first step in the process of mobilization. Although there were Chief Constables, a constabulary of officers for them to command was many years away.

On 3 August 1803 a further letter was circulated from Downing Street to the Lords Lieutenant which appears to be part of a dialogue between the Government and the Lords Lieutenant in which the latter were of the view that more funding was called for from the Government to fund the yeomanry, as distinct from the infantry, than the Government was inclined to provide:

Whilst it would be inexpedient to establish a Volunteer Force to the extent proposed... upon the allowances specified in the printed regulations; and although ... the persons forming new Corps have intimated their intention of putting the Government to no expense, it could not be expected however liberal and public spirited disposition of individuals may be, that such arrangement would become general.

...Volunteer Corps of Infantry ... will be at liberty to draw the following allowances: Twenty Shillings per man for Clothing once in every Three Years and One Shilling per day for Twenty Days exercise within the year provided such allowance be not drawn for any exercise on a Sunday...

I am further to acquaint your Lordship, that no allowance will be made for Clothing and Appointments to any persons who may, from this time, enter into the Yeomanry Corps; but as by the 31st clause of the Act it is provided that any person who shall repair on horseback, properly mounted, armed and accoutred at his own expense, may serve in any Regiment or Corps of Cavalry and shall not be compelled to serve in any Regiment Corps of Infantry, I am to signify His Majesty's pleasure, that the Corps of Yeomanry already established should be permitted to receive individuals so mounted, armed and accoutred into their several Corps or that separate Corps of Yeomanry may be formed of the persons so willing to serve, provided that no additional charge, under the head of contingencies, is brought upon the Government beyond that which is authorised by the existing regulations.

'Existing regulations' provided that:

no greater Expense can be incurred by any Troop of Cavalry ... than One Hundred and Twenty Pounds per Annum to be issued Half-yearly, out of which all Expenses whatever, except for Arms must be defrayed, when the Corps is not called out into actual Service.

In order to save your Lordship the trouble of an unnecessary correspondence it may not be improper for me to apprize you, that the allowances specified in the foregoing part of this communication have been settled upon the most mature consideration, and that I am not aware of any circumstances that would be likely to induce any alterations therein....

Not much room for misunderstanding there, although Colonel Eyre, who had just been elected MP for Nottinghamshire, expressed concern that there would not be enough volunteers who would be sufficiently well off to be able to comply with such conditions for the Yeomanry Troops to recruit up to their full establishment. Once again, however, the incentive of not having to serve in the infantry proved sufficient to ensure that the Yeomanry Troops were recruited fully. Indeed some might offer the view that this powerful incentive has done much to ensure that the Yeomanry has been fully recruited ever since!

In order to meet their obligations to raise additional units, the Lords Lieutenant turned first to their Deputy Lieutenants, most of whom responded by raising at least one sub-unit of either militia or Yeomanry. Lieutenant Colonel Eyre arranged for as many as possible of those troops that had been disbanded the year before to be re-raised, but on a new establishment of a hundred from the previous one of sixty. The following correspondence from the Duke of Portland, in his capacity as Lord Lieutenant, to the Government relates to this.

LONDON
5th August 1803

MY LORD

I have to request Your Lordship to do me the Honour of laying before His Majesty the following offers which I have received from different gentlemen of the County of Nottingham ...

Firstly, Mr Eyre, one of the members of the County and the late Lt.Col. Commandant of the Yeomanry Cavalry raised there in the course of the late war, proposes to raise four Troops of Yeomanry Cavalry upon the terms of an Act passed for the Regulation of Corps of Yeomanry, and I am to observe to Your Lordship that one of the Troops is already complete ...

The Newark, Holme Pierrepont, and Bunny Troops had not disbanded. Not only that but it is said that in the case of Newark not only was the Troop fully recruited but in addition 900 people out of a population of 7,000 volunteered to form a Volunteer Corps. The Troop that was 'already complete' was the Nottingham Troop and the remainder were the Retford, Mansfield and Worksop Troops. The request was duly granted, as was a further request that the Troops could be gazetted when ready and did not have to wait for the others.

The officer establishment on reforming was as follows. The date is the date on which they were gazetted:

Troop	Name	Date Gazetted
Retford Troop	Captain Anthony Hardolph Eyre of Grove	8/8/03
	Lieutenant Samuel Crawley	8/8/03
	Lieutenant Stephen Atkinson	24/08/04
	Cornet Samuel Thortel	27/11/04
	Surgeon John Hull	17/12/03
	Ensign Henry Mason	24/08/04
Mansfield Troop	Captain William Boothby of Edwinstowe	8/8/03
	Lieutenant Francis Vickers	8/8/03
	Cornet James Maltby	8/8/03
Newark Troop	Captain Francis Chaplin	15/8/94
	Lieutenant Phillip Palmer of East Bridgford	15/8/94
	Cornet Francis John Brough of Newark	15/8/94
	Cornet T. Bland	3/4/04
Nottingham Troop	Captain Ichabod Wright of Mapperley	8/8/03
	Lieutenant Alexander Hadden	8/8/03
	Cornet Thomas Wright	8/8/03
	Cornet S. Christopher Colough	3/4/04
	Surgeon Thomas Bland	3/4/04
Holme Pierrepont Troop	Captain the Hon. Charles Arthur Herbert Pierrepont	
Bunny Cavalry	Captain Joseph Boultby	
	Lieutenant William Timm	
	Cornet Henry Breedon	

In addition there was one new Troop, raised by Richard Lumley-Savile of Rufford Abbey, which he called the Rufford Troop. He was the second of three sons of the 4th Earl of Scarbrough, and Viscount Lumley of Waterford, who had married Barbara Savile, when he was aged twenty-seven. She was the only surviving sister of, and heiress to, Sir George Savile, 8th Baronet of Thornhill near Dewsbury in Yorkshire and Rufford in Nottinghamshire. Sir George died in 1782. His antecedents had owned Rufford for generations and had always been one of the leading Nottinghamshire families. Given that the Earl's heir would inherit the Scarbrough family estates at Sandbeck in Yorkshire, close to the boundary with Nottinghamshire, and Lumley Castle in Durham, Sir George left his estates to Richard.

Richard Lumley-Savile was in his early forties, on the old side to command a Troop, but had spent about ten years in the Army, serving successively in the 10th Dragoons, 86th of Foot and the 3rd Dragoons, leaving on inheriting the Rufford Abbey estate. He was MP for a Lincolnshire seat in the mid-to-late 1780s and was appointed Sheriff of Nottinghamshire in 1793. His politics were strongly Tory, being a committed opponent of both Catholic Emancipation and Reform.

He was a Deputy Lieutenant and in pursuance of his obligations as such had decided to raise a Corps of Volunteer Infantry as well as a Troop of Yeomanry. He did not wish the Troop to be independent, but to be part of the Nottinghamshire Yeomanry. He had apparently written to Lieutenant Colonel Eyre to that effect for, on 18 August 1803, Lieutenant Colonel Eyre wrote to him a most revealing letter, partly quoted earlier, giving a clear idea of what was involved in raising a Troop of Yeomanry and also giving an indication of the relationship between the various Troops at the time:

Dear Savile

As you desire me to give you my opinion respecting the clothing and arming of your Troop I will send it without hesitation or apology though if it differed from yours I should do it with the greatest diffidence.

I own I always was of opinion that as little expense as possible ought to be incurred in equipping Corps of Volunteers except in articles that were really useful: and though our Corps has not acted up to this principle yet it is well-known that I only gave way to the wishes of others.

The inconvenience is now materially felt - the compliment, which you were so obliging as to intend to pay our Corps, is now perfectly unnecessary: in the first place I think we shall be obliged to act as perfectly Independent Troops as our members will be so unequal (and as the Worksop Troop is upon the old establishment). And as our uniform is unfortunately expensive it is unreasonable for us to expect other Independent Troops should pay us the compliment of adopting our dress at considerable expense and inconvenience – our jacket alone costs five pounds. I should think you might have one of blue cloth with four rows of buttons very close fitting for half the money.

The price of Swords (at Mr Beaumont's, Birmingham where we are supplied) is as follows:

Swords with polished steel scabbards	32s each
Swords with extra glazed scabbards	29s each
the price of pistols I think it is	25s each

I enquired also the price of flintlocks and Rifles:

Good flintlocks and bayonets	46s each
Rifles	Four guineas

As to accoutrements I apprehend that a Sword Belt would be sufficient for your Cavalry as one of the Holsters might be fitted up to hold cartridges. We shall be able to find arms for our Troops but we shall have more difficulty in procuring Sergeants and Trumpeters. In that respect you will be in the same situation with us.

I confess I think the conduct of Ministers with respect to Volunteer Corps, very unjustifiable. In the first place they officially notify the terms on which Corps of Yeomanry will be accepted and then after many have nearly equipped themselves under the sanction of this notification they withdraw their offers and will grant no allowance at all.

This last paragraph reveals that the art of ministerial moving of the goal posts was well developed, even at that time. The letter continues to complain about 'the system'. Two centuries later do we not from time to time suffer from the same?

In fact the Worksop Troop, which has been mentioned earlier, did not in the end revive. The Worksop Volunteer Association did, however, form a militia, led by the Foljambe family, and the explanation may be that the sub-unit may have re-formed, but as a militia.

The reason for the concern about sergeants and trumpeters was that men of sufficient experience to hold the rank and perform the role of sergeant are key to the operational viability of any military unit, and without trumpeters it was not possible to control cavalry since it was the only effective means of conveying commands in the field; in effect no trumpeter, no cavalry. This remained such a critical matter that the legislation was later altered to allow each Troop Commander to appoint one sergeant and one trumpeter and provide them each with a horse and they became entitled to receive pay for the time they were engaged and could be entitled to be billeted.

The Rufford Troop was duly raised on 21 August 1803. It was an Independent Troop but given that Charles Pierrepont, as he then was, had presented the Royal Standard to the Regiment in 1795 in the name of Captain Lumley-Savile's wife the Lumley-Saviles must have been closely linked to the Regiment.

The Orbat was as follows:

Troop	Name	Date Gazetted
Captain Commandant	The Hon. R. Lumley-Savile	5/9/03
Lieutenant	The Hon. Philip Pierrepont of Thoresby	5/9/03
Cornet	John Vessey, Gent. of Rufford	

Philip Pierrepont was Lord Newark's fourth son, Philip Sydney Pierrepont, who was now aged twenty-one. The Troop numbered sixty all ranks and they came from Wellow, Bilsthorpe, Winkburn, Maplebeck, Kirklington, Hexgrave, Hokerton, Rufford, Boughton, Ollerton, Mansfield, Cuckney, Osmanthorpe, Haughton, Eakring, Laxton and Budby, a considerable area which appears to consist of the Rufford and Thoresby Estates, but, as mentioned earlier, was probably a fairly typical recruiting area for a troop of yeomanry.

Captain Lumley-Savile procured a Letter of Service for the Troop from George III which stated that its terms of service were 'absolutely limited to their Place of Exercise in Rufford Park at all times and to be called from thence in case of actual invasion only'. It seems that the King took a keen interest in the raising of volunteer units of all kinds, because this is the second instance in this account of his direct intervention. Apparently it was acceptable for someone raising a troop of yeomanry to specify its terms of service in this way, which was a similar restriction to that imposed by the Worksop and Bunny Troops. Whether this was an official right or merely pragmatism is not clear. It may just have been the King meddling. The Defence Act certainly did not allow for it.

The Newark, Retford, Mansfield, Rufford, Nottingham, Holme Pierrepont and Bunny Troops totalled 547 effectives altogether. But it should be noted that Lieutenant Colonel Eyre was this time gazetted 'Captain Commandant' and not 'Lieutenant Colonel', thus confirming the fact that the Yeomanry consisted of seven separate and independent troops rather than one unit or 'Corps'.

To give some idea of the scale of the mobilization, by 1804 the following Nottinghamshire-based yeomanry troops and volunteer militias had either been maintained in existence from previous threats or had been raised or re-raised. The name of the commander is given where known:

Name of Unit	Commander	Type of Unit
Bunny Cavalry	Captain Joseph Boultbee	Yeomanry
Bunny (Loyal)	Lieutenant Colonel Cmnd Samuel Wright	Militia
Burton Joyce		Militia
Colwick	Major Cmnd John Musters	Militia
Holme Pierrepont	The Hon. Charles Arthur Herbert Pierrepont	Yeomanry
Loyal Wollaton, Lenton,and Beeston Volunteer Infantry	Lord Middleton	Militia
Mansfield Cavalry	Captain William Boothby	Yeomanry
Newark Cavalry	Captain F Chaplin	Yeomanry
Newark Loyal	Colonel Cmnd Joshua Thoroton Lieutenant Colonel J. Bradshaw	Militia
Nottingham Cavalry with a Company attached	Captain Ichabod Wright	Yeomanry
Nottingham Rangers aka Oxton Volunteer Corps	Lieutenant Colonel Cmnd William Sherbrooke	Militia
Nottingham Loyal	Lieutenant Colonel Cmnd John Elliott	Militia
Nottingham Supplementary Militia		Militia
Ossington	Captain John Kington	Militia
Retford Cavalry	Captain Anthony Hardolph Eyre	Yeomanry
Retford East	Captain Nathaniel March Captain John Parker	Militia
Ruddington Volunteer Infantry		Militia
Rufford Riflemen	Captain John Kirk	Militia

Rufford Cavalry	Captain Cmnd The Hon. Richard Lumley-Savile	Yeomanry
Sherwood Rangers	Captain Sir Thomas Woollaston White Bt. Captain John Shergold	Privately raised and funded infantry
Southwell	Major William Wylde	Militia
Thoresby	The Hon. Charles Arthur Herbert Pierrepont	Militia
Trent Vale	Captain Robert Padley	Militia
Welbeck	Captain William, Marquess of Titchfield	Militia
Wollaton	Lieutenant Colonel Henry The Lord Middleton	Militia
Worksop	Captain F. F. Foljambe	Militia

This list, compiled from more than one source, may not be completely accurate, but it gives a fair idea of just how seriously the people of Nottinghamshire viewed this new threat. Since it was part of a national mobilization in which a similar response was achieved elsewhere, it also shows how seriously the threat was perceived nationally. There are some interesting facts to be gleaned from this list of Nottinghamshire sub-units: note how many of the Militia had 'Loyal' in their title; there is a clear message in that fact which challenges the natural assumption that loyalty to Crown and Country could be taken as read. Clearly it could not then; can it today?

Not all the major estates raised a troop of yeomanry. The majority, in fact, opted for the more economical militia. Several raised both or, in some cases, shared a troop of yeomanry and raised their own militia, the key examples of this being Thoresby and Rufford.

The Pierrepont family already had a Troop at Holme Pierrepont and now helped to raise one at Thoresby. It is assumed that it was Charles Pierrepont, the erstwhile Captain of the *Kingfisher* who took command of the Holme Pierrepont Troop because Evelyn Henry Frederick Pierrepont who had raised the Troop in 1798 had died aged only twenty-six in 1801. However, Charles is also shown in command of the Thoresby Militia, so the position is not absolutely clear. He must have delegated one or other task to someone else, maybe Captain Thomas Bettison who had commanded prior to the pause in hostilities.

The second son had also died in 1800 so, as has been mentioned, Charles Arthur Herbert Pierrepont (1778–1860), to give his full details, was now Lord Newark's heir. No doubt the unexpected deaths of his elder brothers had caused him to cut short his very promising career in the Navy and return to fulfil his new role in life. As a result, when his eldest brother had died in 1801, he had taken over from him as MP for Nottinghamshire and had now in 1803 left the Navy on half pay, giving the whole of the arrears of his half pay to naval charities. In 1804 he married Mary Letitia, Lieutenant Colonel Anthony Hardolph Eyre's eldest daughter.

The reference to Lord Middleton is to Henry Willoughby 6th Baron Middleton (1761–1835) of Wollaton Park, the son of the 5th Baron Middleton mentioned in relation to the presentation of the Standards to the Regiment in 1795. He had succeeded to the title in 1800 on the death of his father. The Willoughbys were a very influential family in Nottinghamshire but originated from Middleton in Warwickshire, being ennobled early in the eighteenth century. They had provided a number of MPs for Nottinghamshire seats and were major landowners not only in Nottinghamshire but in Warwickshire and Yorkshire as well. The family and title is still in being but they gravitated to their estates in Yorkshire where they now live and sold Wollaton Park to the Nottingham Corporation in the early twentieth century.

The Lieutenant Colonel Elliott mentioned as commanding Nottingham Loyal must be the same man as Ichabod Wright is said to have taken over command of the militia from in 1808.

Ossington was, of course, at that time the home of the Denison family who have already been mentioned.

Even an estate as large as the 3rd Duke of Portland's own estate at Welbeck did not raise a troop of yeomanry, although it did, of course, raise a militia. Almost certainly it will have contributed to the Mansfield Troop, however, because there are bound to have been individuals on the Welbeck estate who were willing and able to equip and mount themselves.

The Marquess of Titchfield, mentioned as commander of the Welbeck militia was William Henry Cavendish-Bentinck (1768–1854), the second son and heir to the 3rd Duke of Portland. He was educated mainly in England but, presumably in acknowledgement of his Dutch ancestry, completed his education at the Hague. In 1790, he was elected MP for the Borough of Petersfield but, in 1791, exchanged this constituency to become Knight of the Shire for Buckinghamshire, which enabled him to sit in five successive parliaments without having to contest an election.

Clumber, the seat of the 4th Duke of Newcastle, seems to have raised nothing at all, but presumably continued to contribute to the Retford Troop. At first sight this is a significant omission, given that the 4th Duke was destined to take command of the Regiment in only eight years' time. There are two reasons why the 4th Duke was not involved at this time: the first is that he was still only eighteen years old and the second is that he could hardly raise a Troop, since at the time he was a prisoner of the French. He, aged seventeen, having completed his education at Eton, had, like many others, taken the opportunity afforded him by the peace created by the Treaty of Amiens to venture on a continental tour. There was a keen interest amongst the British to view the sites of the Revolution. Unfortunately, whilst he was in France, hostilities were renewed and he had found himself detained at Tours. The irony of this tale is that his coat of arms includes 'two demi-belts, with buckles arg. Erect', as an honorary augmentation in memory of his ancestor Sir John Pelham who had taken prisoner John, King of France.

Last, but not least it can be seen that the dedicated Sir Thomas White had re-raised the Sherwood Rangers, still entirely at his own expense.

Chapter 9

On the Brink of Invasion 1803

The role for which the Yeomanry were to be trained was to be capable, when ordered, of being called out and assembled

> for actual Service on any Invasion and or Appearance of an Enemy in Force upon the Coast or to suppress any Rebellion and or Insurrection during any such Invasion.

Troops of Yeomanry were not subject to the compulsory sections of the Defence Acts except in the foregoing circumstances. Note that the emphasis had changed from the wording used in 1794 which mostly related to riot and civil unrest to be primarily concerned with 'invasion' and there was no provision at this stage for call out of the Yeomanry in the event of any 'rebellion and or Insurrection' which was not related to an invasion. This change again reflects clearly the very different nature of this new threat.

By early 1804, despite the feverish activity which has been described as it affected the reserves, the overall situation was still serious. Britain had no allies and the Government was in turmoil and was then defeated and fell. This led to the recall of Pitt. Notwithstanding these changes, from a local point of view, the 3rd Duke of Portland continued to serve in the Cabinet and so nothing fundamental had changed. On the Regiment's front there were many additional signs that this time the Government intended to be much more professional in the way that it prepared these yeomanry regiments. It was seeking much better levels of capability than previously. Before long the Government was ordering all ranks to perform fourteen days' duty annually, officers and men being billeted for the duration. For this they were paid. This was the first time that they had been offered formal pay, as opposed to the levy. The annual period of continuous training has remained in force almost continuously ever since, although for many years it was reduced from fourteen to only eight days' annual duty.

Some clear orders were issued, for example in March 1804 an instruction was issued from Horse Guards, the headquarters of the Army, as follows:

INSTRUCTIONS to be observed by the Volunteers if suddenly ordered to March and assemble on any named point.

Being provided with Horses and Carriages as pointed out in the Secretary of State's Letter of the 16th January, the Corps will leave every Incumbrance whatever behind them. No Women are to move with the Corps on any Account. They will March as much as possible together with the Brigades or Corps into which they are formed, agreeable to the particular Routes and Orders that had been forwarded from the Headquarters of their District, to be acted upon after the Certainty of the Enemy's appearance on the Coast.

In the first Hurry of assembling the Troops on the landing of an Enemy, it may be impossible (in all Cases) to prescribe positive Routes and to prevent crossing, crowding, and interference in the March of so many Bodies moving from distant Places and tending to the same Point. The Prudence and Arrangement of the Commanding Officers must, therefore, as far as possible, provide against these unavoidable Difficulties.

When the Troops marched in Corps and encamp they are more easily provided for. But when they are cantoned and pressed forward they must be satisfied with the most crowded and slightest Accommodation and a rigorous Discipline must prevent them from requiring more than the Country and Circumstances can allow.

On the Routes where Troops are to March sufficient Bread and Oats may certainly be provided by the Army Bakers, and Purveyors of Oats in the Market Towns on a short Notice: should not this be the case Commanding Officers must themselves make the best General Arrangements on this Head that Circumstances will permit and not allow Individuals to trespass or permit Outrages: and it is again earnestly recommended to the Corps both of Cavalry and Infantry also to the Proprietors of Waggons which are employed for conveying Troops to leave their Homes with not less than Three Days Bread and Forage.

As soon as the Corps collect in sufficient Bodies and the Troops canton or encamp the Commissariat (whom every necessary Assistance must be given) will provide for them.

If there is no Commissary and that the Pressure is such that Provision of Straw, Forage, Bread, Fuel must be provided from the Country, the Commanding Officer will appoint one or more intelligent Officers to act as Commissaries for the Time being, who will go forward with proper Assistants, and call, in the first Instance, upon the Commissariat Contractor, of which there has been one appointed in each Market Town, to furnish the Supplies required; but in Situations where there is no such Contractor or that he is unable to provide for the Demand, the Officer, is to address the Magistrate or Magistrates of the Neighbourhood and representing the Necessity of the Case induce them to take such Measures and give such orders as will with due Attention to the Preservation of Property and without Interference with the Discipline of Troops immediately produce the necessary Supplies for all which distinct Requisitions in Writing signed by the Commissary (or Officer acting is such) must be presented specifying the number of Persons and Horses and Quality and Rate of Rations and the Number of Rations and for what Time. Upon receiving the Supplies proper Receipts signed by the Quartermaster or other Person must also be given according to accompanying Forms.

When Troops encamp they will chuse [sic] the Commons where there are such, otherwise they must encamp on such Grass Fields as are most convenient for their March and to which they can do the least Damage and if unavoidable Damage in such case is done, the Commanding Officer of the encamping Corps will grant a Certificate of the Nature and Amount.

Another key part of the preparations was the decision promulgated on 2 August 1804 that the country was to be divided into separate military districts and those yeomanries in each district were to be grouped in brigades under command of regular brigade commanders. All the Nottinghamshire Troops were brigaded with others in the Brigade set up in the Northern Inland District which covered Derbyshire, Warwickshire, Staffordshire, Leicestershire, Rutland and Nottinghamshire. The Nottinghamshire Troops formed the left-hand three Squadrons in the Second Line of two in the order of battle of the Brigade. They were paired from right to left as follows:

Lutterworth
Bunny

Nottingham
Holme Pierrepoint

Mansfield
Rufford

Retford
Newark

The Brigade Headquarters had deliberately grouped them geographically within the order of battle rather than by seniority, so that they could train together. This showed that they were indeed regarded by the chain of command as independent troops and regimental affiliations were not officially recognized. A good example of this was the pairing of Bunny in Nottinghamshire with Lutterworth in Leicestershire. However, as a result of this structure a much greater sense of urgency and professionalism began to emerge.

The Brigadier General appointed to command the Brigade was Brigadier General F. Erskine. In addition 'Inspecting Field Officers to the Yeomanry and Volunteers' were appointed throughout Great Britain to assist the Brigade Commanders. Colonel Kane was appointed to support Brigadier General Erskine but his background is not known. This may be the first manifestation of the concept which is now widespread of appointing senior TA Officers as Deputy Commanders of Brigades and similar officers to advise the regular chain of command under which the TA has operated for many years.

Initially the restriction on the deployment of the Rufford Troop did cause difficulty when the Inspecting Field Officer to the Yeomanry and Volunteers with responsibility for the Northern Inland District or Brigadier General Erskine himself tried from time to time to summon the Troop for its annual inspection to a location of his own choosing. Initially Captain Lumley-Savile declined graciously. He obviously did not do so lightly. The rough drafts of his letters survive and the many amendments bear witness to the care taken over them. His basic point was that his Troop was widely dispersed. It was therefore unreasonable to expect them to both assemble at Rufford and then to undertake a significant march to another location and then back home in a single day. In the end Brigadier General Erskine issued a direct order declaring null and void the previous concession and thereafter the arrangements for the Rufford Troop were the same as for everyone else.

All other Troops in Nottinghamshire were obliged to extend their terms of service to include the whole of the North Inland District and were consequently liable be called out by the General commanding the District whenever he thought proper.

On 2 October 1804 Brigadier General Erskine issued the following Brigade Order:

Brigade Order

Brigadier General Erskine having finished the Inspection of the Brigade under his Command is highly satisfied with the general appearance of the Men and Horses and the General Good State of the Arms Accoutrements Etc. and feels confident that should their Services be required so respectable a Body of Cavalry must provide a most important Arm in the Defence of the Country. He at the same time begs to impress in the strongest manner on the minds of officers commanding the absolute necessity of every Corps being drilled on the same principal and the words of command the same as laid down in the Instructions issued by command of His Majesty.

Extensive detailed instructions were then issued in writing, a copy of which survives and is in the Regiment's archives, giving every conceivable word of command in every conceivable situation.

On 22 December 1804 the Ordnance Office issued a District Order stating:

the Master General and Board of Ordnance having ordered that in future the proportion of Ammunition which is established as the usual allowance for the Exercise and Practice of Volunteer Corps

In Gentlemen and Yeomanry Fencible Volunteer Cavalry:

Ball and Cartridges: 10 per man
Blank ditto: 24 per man

An instruction relating to the pay of Yeomen was issued from Whitehall on 1 April 1805:

The Secretary at War will be guided in the Issue of the Pay and Allowances, and the Receiver General of the County authorised to pay the Marching Guinea for any Period of Duty not less than 21 Days (or the Proportion thereof, at the Rate of One Shilling per Diem if for less than that Number) by the Return transmitted through the Lord Lieutenant of the County to the Secretary of State.

Two days before that instruction was issued, on 30 March 1805, the opening move took place in a sequence of events that would end in permanent relief to Great Britain from the threat of invasion by Napoleon, or anyone else, for well over a century, and immortality for Nelson.

For the previous two years Nelson had been at sea in the Mediterranean blockading Villeneuve, the best admiral the French had, in Toulon. Suddenly Villeneuve's fleet slipped to sea past Nelson's fleet whilst Nelson's attention was momentarily elsewhere. By the time Nelson had recovered and sailed in pursuit, Villeneuve was through the Straits of Gibraltar. As Nelson entered the Atlantic, two weeks behind Villeneuve, he analyzed the scraps of information available to him and intuitively took the decision to sail for the West Indies, guessing correctly as it turned out, that this was Villeneuve's destination. Villeneuve's plan in crossing the Atlantic was to make a feint towards the Caribbean with the intention of deceiving Nelson and then to turn back leaving Nelson out of position, drive past Admiral Cornwallis, whose fleet, the Channel Fleet, was guarding the Western Approaches, into the Channel and thus secure naval supremacy in the Channel for long enough for Napoleon to invade England.

Nelson completed his crossing on 4 June and Villeneuve, on being informed of his arrival, and unknown to Nelson, immediately began to re-cross. Nelson reached Antigua and found that Villeneuve had been there only four days before. He deduced that Villeneuve was on course for the Channel and sailed in pursuit sending the *Curieux*, a fast sloop, ahead, tasked with overtaking Villeneuve to warn the British Government.

The sloop dropped anchor in Plymouth on 8 July, having spotted the French fleet as she passed them to windward and thus safe from pursuit. By the 9th her captain, George Bettesworth, had reported to the First Sea Lord at the Admiralty that Nelson had set a course to the south of Villeneuve in order to cut him off at Cadiz from re-entering the Mediterranean. He also reported that Nelson had made the appreciation that Villeneuve's destination was the Channel. In fact, although Nelson was correct, Villeneuve's first port of call was to Cape Finisterre. His purpose was to strengthen his fleet by releasing another French squadron and his ally the Spanish Fleet blockaded there in Ferrol and Corunna.

The Admiralty moved with a speed that Napoleon, when he heard of it, could not credit, so that by 11 July Admiral Cornwallis had orders to hand to reinforce the Squadron blockading Ferrol and Corunna, which was under the command of Vice Admiral Sir Robert Calder, with that commanded by Rear Admiral Stirling off Rochefort. Admiral Cornwallis himself, whose orders were to deny Villeneuve access to the Channel, was ordered to patrol in particular between Ushant and Finisterre. Calder, even when reinforced by Stirling, was outnumbered by the force which faced him which now consisted of the whole of the French and Spanish fleets.

Villeneuve, with the intention of forcing access to the Channel, put to sea and, on 22 July, closed with Calder under cover of fog. The engagement, which became known as the Battle of Cape Finisterre, lasted four hours and was inconclusive, the only definite outcome being the capture by Calder of two Spanish ships. Calder broke off the action. He offered action on the next two days but Villeneuve declined, taking refuge in Ferrol. This action was mired in controversy. The Admiralty was very critical of Calder for not having forced an all-out engagement, despite his numerical inferiority, his failure meaning that the French and Spanish Navies were still at large and a continuing threat, because they were the one way in which Napoleon could mount an invasion. Napoleon's own view subsequently was that the engagement had effectively ended his chances. Napoleon continued to believe that the decisive naval action of the war was this rather than Trafalgar.

Nelson, although in overall command, at this point took HMS *Victory* into Portsmouth, so as, on dry land once more, he could report to the Admiralty.

Napoleon, not at that point realizing the extent to which Villeneuve had been outfaced and therefore expecting him in the Channel daily, now stood to his army of 200,000 in number and began his final preparations for the invasion. So apparent was it that he was about to make his move that Horse Guards issued warning orders nationally as a result of which, on 11 August 1805, the Regiment in turn received the following order from Brigadier General Erskine:

in consequence of Government having received Intelligence which is confirmed that large bodies of troops are embarking on the Coast of Holland and that the Enemy have given Symptoms of increasing Activity at Boulogne and its Neighbourhood I am directed to inform You of the Possibility of the Yeomanry and Volunteer Force being speedily called upon for service.

The critical moment was now at hand.

On 28 August, the following Brigade Orders were issued to the Regiment:

Brigade Orders

In the event of the Brigade of Yeomanry Cavalry of the North Inland District, being called into actual service, or ordered out for Inspection, or any other Purpose; it is highly proper that the Yeomen should be made acquainted with what necessities are essential to their comfort upon all such occasions. — They are therefore required at all times to turn out with the following Articles, (and no more), in their Cloak Cases or necessary Bags.

Two Shirts - One pair of Worsted Stockings - one pair of strong Shoes - One Comb - One Cloaths and two Shoe Brushes - One Brush, and piece of spunge for pipe Clay - A Shaving Case - a Watering Cap - One pair of Ruffia Duck Trousers, if not provided with Overalls - one Curry Comb and Brush to be packed under the flap of the Bag with the handle out.
A Stable Jacket to be neatly folded and carried above the necessary Bag over which a Blanket is to be carried, likewise carefully folded, the whole covered with a water Deck.

The feeding Bag and Snaffle Bridle, to be fastened to the ring of the Cantle of the Saddle, offside, likewise the forage cord when not in use.

The Corn Bag will be carried across the Saddle – the Canteen and Haversack, to be flung over the right Shoulder.

The following of the above-mentioned Articles, will be furnished by Government, when the Brigade is ordered to March.

Canteens - Haversacks - Feeding Bags - Corn Sacks - Forage Cords - Water Decks - Likewise Camp Kettles - Water Buckets - Bat Saddles, Picket Posts, and Breast Lines

Each Man to provide himself immediately with two fore foot shoes and a Picker, which he will carry in the shoe case.

One Bat Horse to Carry the Men's Camp Kettles, and one for the Officers Baggage, must be attached to every Troop.

Commanding Officers of Regiments and Officers Commanding Troops will occasionally accustom their Men to pack and carry their necessaries in a neat and Soldier-like manner; which besides adding to their appearance, must be regarded as contributing to their personal ease, and still more to that of their Horse – the Horse Cloth to be laid aside.

Extracts from the General Instructions

Carriages allowed to be with each Regiment of Cavalry of 10 Troops, at 60 Rank and File, and upwards.

2 Bread Waggons (Each to carry at the rate of three days Bread for 500 Men)
2 Waggons for two days Oats for each 300 horses
1 Ammunition Waggon
2 Forge Carts
1 Waggon for Sick
1 Suttlers Cart

Bat Horses independent of those mentioned for each Troop.

1 for Entrenching Tools
1 for Surgeons Medicines
6 for the Field Officers and Regimental Staff

Regiments on lower establishments will be allowed Carriages and Bat Horses, in proportion to their effective strength.

Also on 28 August Lieutenant Colonel Eyre wrote to all the Troop Commanders at the instigation of Brigadier General Erskine to propose the appointment of one of their number as commanding officer. A meeting was called but the result is not specifically known. Presumably, however, it was Lieutenant Colonel Eyre who was chosen. It is assumed the reason for this move was that Brigadier

General Erskine realized that if his Brigade was actually to go into action he would need his sub-units to adopt conventional regimental groupings to aid command and control.

Meanwhile, on 14 August, Villeneuve had sailed once more for the Channel only to discover the British fleet under Cornwallis blocking his path. Villeneuve had realized that to risk a battle would be suicidal and had broken off contact towards the south, reporting to Napoleon that it was beyond the capability of the combined French and Spanish Fleets to control the Channel for long enough to enable an invasion to take place.

So as the Regiment had prepared to march to meet Napoleon's imminently invading army, in reality Napoleon, in recognition of the invincibility of Nelson's Fleet, had ordered his army to strike its tents and march away from Boulogne for the Danube. Early September brought despatches to London revealing that not only had Napoleon marched away but Villeneuve had retreated south. With the French now on the back foot at sea, Nelson sailed from Portsmouth once more and went in pursuit of Villeneuve, determined to drive home his advantage ruthlessly. Villeneuve was making for Cadiz where Nelson trapped him and prepared to wait for him to put to sea. Nelson would not have to wait long because as September 1805 ended it gave way to one of the most momentous months in the history of Europe.

October opened with the French despatching south a replacement for Villeneuve, nominated because of Villeneuve's failure to face Nelson in battle. This was followed shortly thereafter with Napoleon's massive victory in Austria at the battle of Austerlitz where he smashed the Austrian and Russian alliance so carefully constructed by Pitt's diplomacy. Next Villeneuve, thinking it better to be branded a dead hero than a live coward put to sea before his replacement arrived and suffered his mighty defeat at Trafalgar on 21 October through the tactical genius of Nelson who was, of course, mortally wounded. Finally, as the month ended, there arrived in England from service in India a highly regarded young general on the fast track to promotion. His name was Arthur Wellesley, later Duke of Wellington.

Although Nelson's victory at Trafalgar largely brought an end to the threat, which the yeomanry and others had been re-raised to meet in 1803, it did not bring about a change in the country's basic situation. Whilst the Navy emphatically ruled the waves a state of war continued between the British and the French in which the British were without allies against a France that now controlled almost all of Europe with the exception of Austria, Russia, Prussia and Portugal. This rendered the continent, both militarily and economically, pretty much a no-go area to the British. In addition, at home, whilst the threat from republicanism had reduced substantially, new threats to social stability were emerging, arising out of the first stirrings of the industrial revolution, of which more later. The tensions created in Parliament by these burdens, on top of Napoleon's victory at Austerlitz, finally broke Pitt who died in 1806.

He was replaced initially by Lord Grenville who led a coalition called the Ministry of All the Talents, mostly consisting of Whigs, in power for the first time since 1783 and the last time until 1830. Hopelessly out of touch with government, they lasted little more than a year, being replaced in 1807 by an administration of largely Tory complexion, led, on a caretaker basis, by the 3rd Duke of Portland. Both governments had in common a commitment to the war but a lack of grip so far as its prosecution was concerned, military effort being frittered away on ineffective forays along the Mediterranean coastline. However, where the 3rd Duke of Portland made a crucial contribution was in blooding a new generation of national leaders trained by Pitt, the most prominent among whom were George Canning and Viscount Castlereagh.

Nevertheless, from the watershed created by the events of August 1805, a gradual reduction in tension ensued for the time being. This was because one of the consequences for the French in being unable to invade England was that England, through its Navy, was able to prevent Napoleon's mighty Empire from gaining access to any external markets or even from maintaining diplomatic relationships with any other country. So, once again, since it suited his intention to open these channels, Napoleon encouraged an easing of the tension between the two countries to such an extent that in 1806 the French released a number of detainees, including the 4th Duke of Newcastle, now aged twenty-one, who returned home.

In 1807 the continuation of the Rufford Troop was placed in doubt because Richard Lumley-Savile, its Troop Commander, succeeded as 6th Earl of Scarbrough, on the death of his elder brother without issue. Under the terms of the will of his uncle, Sir George Savile, through whom he had inherited the Rufford Abbey estate, he now forfeited his right to it to the youngest of the three brothers, The Reverend and Hon. John Lumley-Savile, and therefore ceased to have any interests in Nottinghamshire, a significant loss to the county in the affairs of which he had played such a full part. Unfortunately his younger brother, despite having taken Holy Orders in 1785, turned out to be a disagreeable and litigious gentleman, better known as 'Black Jack', who duly took up residence at Rufford.

The relatively minor part played by the Regiment, in common with many others throughout the land, in the Napoleonic Wars was now at an end. The fact that no invasion materialized and therefore the Regiment never saw active service at this time and in this role should not obscure the fact that, during the period between 1794 and 1805, a span of eleven years, those directly involved regarded the threat posed first by the French Revolution and then by Napoleon to be extremely severe. At the core of all their actions in raising and training their Troops was a firm conviction that they could well be used in defence of their beliefs against a formidable enemy whose intent was to invade, defeat and occupy this country and fundamentally change our way of life.

Chapter 10

The Peninsular Campaign 1807–15

A campaign on land now began in earnest which culminated in the famous victory at Waterloo in 1815. It was of such significance to everything that occurred in the country both during that period and for the rest of the century that it is worth recounting in outline for that reason alone. However, as will be seen, it also had a significant, if indirect, effect on the Regiment and its future.

In 1807 Napoleon inflicted a crucial defeat on Prussia at Jena and went on to defeat the Russian army on the Eylan river before making an alliance with the Russians. Taking into account that Austria was already a pliant satellite, France now effectively ruled all of mainland Europe, apart from Spain and Portugal.

The 3rd Duke of Portland's leading Ministers at this stage were his protégés George Canning and Lord Castlereagh. They saw this new situation as creating a further threat of invasion by Napoleon, this time from the direction of the Baltic and assessed that, if they did not act, Napoleon would shortly seize control of the very significant Danish navy and use it for that purpose.

Canning and Castlereagh had inherited none of Pitt's diffidence concerning the war and used the Royal Navy in a surprise attack to seize the entire Danish fleet before Napoleon had time to do so himself. This was a brilliant pre-emptive strike, although one which, considering Britain was not at war with the Danes, and omitted to declare war before acting, was not entirely in line with the diplomatic conventions of the day, but these were not ordinary times. This event finally and conclusively removed all chance of a French invasion.

At last realizing that Napoleon could not be trusted and would stop at nothing to pursue his imperialistic ambitions, the 3rd Duke's Government now set about waging war with vigour against him on land and sea. Their next move was to use the Navy to re-impose a vice-like blockade on Napoleonic Europe thus cutting off, absolutely, Napoleon's embryonic trade with the rest of the world. There was only one response possible from Napoleon, the resumption of an all-out state of war against Britain.

Because the British could still trade with Europe through the Iberian Peninsula, Napoleon next turned his attention on Spain, in order to prevent this. His solution was to place one of his sons on the Spanish throne. This triggered a spontaneous popular uprising by the Spanish, the opposite reaction to that which Napoleon had intended, which resulted in the surrender of the 22,000 French troops under General Dupont stationed in Spain. For the first time the British Government was in a position to respond to an opportunity such as that which now presented itself in Spain. Since the seizing of the Danish fleet had removed the last vestige of the risk of invasion, the Government had available to it large numbers of volunteer troops no longer required for home defence which it now mobilized for service abroad. At a local level this mobilization did not stop the gradual retrenchment of yeomanry troops, because they could not, of course, be used overseas.

The mobilization was based on the militia, leaving the yeomanry to provide home defence and to be used in aid of the civil power, a reversion to the role for which it had been raised originally in 1794. There was a continuing need for local troops of some sort and of a certain size, to deal with local civil unrest which, as will be seen, was now a major issue. The concern with using the militia, rather than the yeomanry, for policing that threat, was that it meant 'putting arms in the hands of the most powerfully disaffected'. In the unstable times which now existed, this was an unacceptable risk.

A number of proposals were made to facilitate mobilization at this time. The one that had been favoured by the 'Ministry of All the Talents' was the abolition of all the militia and their replacement by the plan set out in the Training Act proposing the call up of the entire population of serviceable age for service with the colours in batches of 200,000. This legislation was repealed by the 3rd Duke of Portland's government who restored the militia but then mobilized it nevertheless into regular regiments for war service. It is hard to know just how calculated were these moves. It was certainly convenient that one of the results of the escalation of the war with Napoleon was that the nation's young hotheads, 'the most powerfully disaffected', were overseas, away from where they could make trouble, in domestically troubled times, and could only point the 'arms in their hands' at the French, which they learned to do with considerable success. Whilst these individuals were replaced with recruitments to raise more militias for home service it is reasonable to assume, the 'cream' of youth having already been enlisted, that these were older and steadier.

The Sherwood Rangers had played their part in the response to the threat from invasion under the continuing leadership of Sir Thomas White. However, there is only one record of their being stood to. This was shortly after the French had broken camp at Boulogne and, therefore, when the threat of invasion was much reduced. The Government decided to test in some parts of the country the popular feeling, and to see whether there would be a general rising of the people in case of an invasion if they were to light the beacons they had set up for that purpose. They selected only certain ones. One of them was that at

Gringley-on-the-Hill near Gainsborough, on the Duke of Portland's estate. Although some ten miles distant it was this beacon from which the Sherwood Rangers received the warning to mobilize. It could only be seen from an elevated position at Wallingwells, so Sir Thomas had arranged for a platform to be erected in a large oak tree at the side of the drive to Langold on which a sentry was posted by day and night to watch for the Gringley Beacon. Unfortunately there was one on duty when it was lit in the middle of the night, and Sir Thomas had not been told that it was a practice only, so he mobilized his unit.

His system for doing this was to keep two horses permanently saddled in his stables and so he was able to assemble in quick time and moved briskly off to Doncaster, their place of rendezvous, some twelve or more miles distant. On arrival, presumably without much sleep, Sir Thomas learned of the false alarm. It is said that on being so informed he 'laughed, and proposed the health of the King in a round of drinks', just as soldiers the world over are wont to do when sent on a wild goose chase by their superiors. He may also have been relieved that it had turned out to be a false alarm since he generously rewarded each man with breakfast and a guinea after which he dismissed them to their homes.

There is no doubt that the unit maintained strong affiliations to the concept of light infantry which was being developed by Sir John Moore throughout this time, and it may well be that it was a source of recruits for the units of the light brigade, later to become the light division, as they were brought up to war establishment.

With the threat of invasion now finished there is strong evidence, despite the lack of detailed records, to suggest that Sir Thomas took the decision to discontinue his heavy financial commitment to maintain the Sherwood Rangers. Rather than disband, however, he appears to have used the unit to form the nucleus of one of the new militias raised for home defence referred to earlier, which would mean that the cost would thereafter be borne by the Government. The evidence for this is that, in 1808, Sir Thomas is known to have taken command of the 3rd Nottinghamshire Regiment of Local Militia, based in Retford and which incorporated the Sherwood Ranger's horn and lanyard as an insignia on its uniform, a command which he held, probably until the Government disbanded that type of militia in 1816 following Waterloo. He also eventually recruited two of his sons into the unit. He had seven children, among them Thomas born in 1801 and Taylor in 1805. Thomas joined as an ensign in 1813, when only twelve, and Taylor in the same rank in 1815 when only ten. Both probably also served till 1816 and attained the rank of lieutenant. Both were destined to serve in the Regiment, the former to command it.

The young Thomas had inherited his father's love of falconry which is said to have led to a lucky escape when he was fourteen. He was on holiday in Scarborough and had set out along the sands to meet his father, who was shooting seagulls with a party of friends. He had a hawk on his wrist. A fog came down and young Thomas turned back, but found himself cut off by the incoming tide and was swept off a rock by a wave. To save himself and his hawk he turned and floated on his back as he had been taught, raising the hand on which the hawk sat. As the tide carried him out, the bird started to flap its wings and, so it is said, with this help young Thomas was able to swim and scramble onto a rock which was just above water level and from which he was later rescued.

The British duly seized on Dupont's reverses in Spain as a pretext, and in 1808 landed 30,000 well trained and well equipped troops in Portugal to assist the Spanish uprising and secure Portugal from the French. One of the senior officers posted to Portugal was Sir Arthur Wellesley. As has been mentioned he had already earned a significant military reputation in India and was, in addition, now a Member of Parliament [he had already been an MP in the Irish Parliament]. The military action which followed did not go well, for which the leadership was blamed, but Wellesley avoided the criticism made of his immediate superiors; indeed, in contrast, his reputation was actually enhanced.

Napoleon reacted to the surrender by his troops in Spain and the landings of the English in Portugal with customary understatement, by marching 250,000 troops into Spain with himself in command. By now the leadership of the British in Portugal had passed from the failed commanders of the initial expedition to General Sir John Moore, who with a much smaller force than Napoleon's, was left with limited options in relation to this new turn of events. Serving under Moore and about to have his first taste of action in the coming campaign, was one Thomas Wildman (1787–1859) who was destined to command the Sherwood Rangers Yeomanry. He was then aged twenty-one and had been educated at Harrow where he became a friend of the poet Lord Byron, whose seat was at Newstead Abbey in Nottinghamshire. It was through this friendship that Thomas Wildman's links with Nottinghamshire eventually arose. In 1808 Thomas Wildman had purchased a cornetcy in the 9th Light Dragoons and later the same year was promoted lieutenant without purchase, and transferred into the 7th Hussars.

Lord Byron was not any ordinary friend. He had a dress sense reflecting his personality which can only be described as extravagant, ostentatious and exotic. Over the years, through numerous thinly disguised relationships, he became widely regarded as bisexual and, through his poetry, which became the height of fashion, was destined to become one of the most charismatic, desired and scandalous personalities of his own or any generation. Somewhere along the way he managed to have himself described by a lady as 'mad, bad and dangerous to know'.

In a brilliant move Sir John Moore now cut Napoleon's supply lines which thwarted Napoleon and left him with little choice but to try and commit Moore to battle. Using a tactic which was to become characteristic of the whole Peninsular Campaign, Napoleon made a forced march by infantry through the open country which abounded in that part of the world. He sought to achieve the classic effect of shock action, namely surprise, hitherto the sole preserve of cavalry and, also, cut Sir John Moore off from his secure base on the coast at Corunna. It was, of course, for exactly this type of open warfare that Sir John Moore had developed the Light Brigade, which now formed an important element of the troops under his command.

Napoleon probably did not realize he was seeking to play Moore at his own game. Sir John immediately appreciated what Napoleon was trying to do and force marched for his bridgehead at Corunna, reaching there before Napoleon. He then turned and secured his defensive line. In the subsequent fiercely-fought battle, and despite being heavily outnumbered, the British held the French. This was a crucial action because it both proved Sir John Moore's concept and restored the reputation of the British Army. Tragically in the hour of his greatest achievement Sir John was killed.

Despite the fact that the French had been held, the Government was unable to capitalize on this success, because they were unable to re-supply and sustain the force in such a precarious position, and so they had no choice but to arrange for its evacuation, leaving the French in control of Spain and only a small British garrison in Lisbon, which would nevertheless prove crucial.

In 1809 the action was mostly elsewhere, the Austrians again posing a threat to Napoleon who, therefore, suppressed them once more, this time at the battle of Wagram following a brilliant campaign involving an Army of 400,000. Meanwhile the British were diverted by an ill-founded operation to capture Antwerp. This operation went so badly that, unbelievably, it resulted in a duel being fought between George Canning, and Lord Castlereagh and the resignation not only of both of them, but of the 3rd Duke of Portland as well, who was replaced as Prime Minister by Perceval.

The one positive move that year, from the British point of view, was that Sir Arthur Wellesley was appointed to command the now-to-be-reinforced British garrison in Lisbon mentioned earlier, resigning his seat in the House of Commons to do so. Although appointed by the 3rd Duke, it was with the approval of Perceval who gave Wellesley his full support once he became Prime Minister and who for his own part then pursued the war with determination.

Wellesley's first move was to try and regain the initiative from the French which he did by striking deep into Spain by means of an impressive forced march by Sir John Moore's legacy, the Light Brigade, already showing signs of being recognized as one of the successes of the Peninsular Campaign. This move enabled him to defeat the French at Talavera. Wellesley realized that, following the battle, he would be too exposed and, therefore, would need to withdraw, which he did, back through Portugal into Lisbon. He was, as a consequence of this short but crucial campaign, created Viscount Wellington.

Wellington, as he was now known, realized that the next move by the French would be to seek to push his much smaller force back into the sea, as they had done with every other attempt by the British to secure a permanent bridgehead in mainland Europe. So Wellington, demonstrating that he was equally skilled in the arts of defence, built a series of parallel defended lines across the Peninsula at the end of which lay Lisbon. These became known as the lines of Torres Vedras. Furthermore, he laid waste the countryside on the enemy side for miles beyond.

Secure behind his defences, well supplied from the sea, and entertaining himself and his troops with various diversions, even foxhunting, he sat and waited for the French. During this period Lord Byron passed through Lisbon, the only gateway into Europe available to him on his way to the adventures further east which inspired much of his most fashionable poetry. No doubt he and Thomas Wildman, who was also there, renewed their friendship.

During 1810 the French sought to invest Wellington's defences. They found that, as a result of Wellington's carefully laid plan, it was they who were starving for lack of local forage and not the British. As a result, the French were forced not only to withdraw from the lines of Torres Vedras, but to withdraw from Portugal altogether.

Wellington was now in a position to take the offensive. His ultimate objective was to drive the French from Spain, but given that his army of 30,000 faced 250,000 Frenchmen in Spain, 100,000 of them directly opposite him, clearly this would need careful planning and then time. It would also need the patience of the Government despite the fact that its appetite for success had now been whetted. By 1811 Wellington, well supported by Perceval, was ready to make his move and began a campaign based on manoeuvre in which the Light Brigade, now the Light Division, played a key part. The opening engagements were at Fuentes d'Onoro and Albuera, neither of which were decisive but left the British in a stronger position.

Although it is a matter for conjecture whether the Sherwood Rangers supplied recruits to the regiments of the Light Division, what is a fact is that Sir Thomas encouraged a member of his family to take a commission in it. This was his cousin Thomas Taylor Worsley, whose widowed mother lived with Sir Thomas at Wallingwells. Thomas Worsley was born in 1794 and so by 1812 he was eighteen years old and would no doubt have, like his cousins, served in a junior capacity in the Sherwood Rangers and then in the 3rd Nottinghamshire Militia before he was commissioned. He was also destined to serve in the Regiment. He was a big man standing 6 feet 4 inches and was commissioned into the 95th (Rifle) Regiment, formerly known as the Rifle Corps. His commission was probably purchased by Sir Thomas since the fact that his mother lived with Sir Thomas indicates that she would have been unlikely to have the means to purchase one for him herself. He travelled out to join his regiment in late-1811, at the age of seventeen.

He was there in time to take part in the great successes of 1812. Two fortified towns lay athwart Wellington's route into Spain and would have to be taken before Wellington could begin to bring the French to a decisive battle. They were Cuidad Rodrigo, Thomas Worsley's first taste of action, and Badajoz which had to be stormed in some of the bloodiest fighting of the whole campaign, only falling at the third attempt. At Badajoz Thomas Worsley had what he described as a stroke of luck; he had been concerned that his height would make him a target, and so it may have done for he was badly wounded whilst in the thick of the fighting when a musketball struck him in the front of his neck, passing round inside the skin, and exiting at the back. He recounted that the stroke of good fortune involved was that had he not been so tall the shot would have hit him in the head and that would have been fatal. He was, however, left with his head at a permanent tilt to one side.

Wellington was now free to manoeuvre inside Spain itself and quickly brought Marmont to battle and defeated him at Salamanca, in time for which Thomas Worsley, complete with tilted head, was back in the line. The victory brought about the fall of Madrid amongst scenes of wild rejoicing. However, with Soult advancing up from the south with an army double the size of his own, Wellington took the only option available and withdrew back into Portugal for the winter.

What a winter did that of 1812–13 turn out to be. First, in England, Perceval was assassinated by a madman and succeeded as Prime Minister by Lord Liverpool, the son of Charles Jenkinson, who had been the organizer of Government patronage during the reign of George III and a close colleague of Pitt the Younger. He was destined to hold office for fifteen years till 1827. Second, Napoleon, having fallen out with the Russians, embarked on an invasion with the aim of taking Moscow. Despite reaching the gates of the city, he failed to take it, and then had to retreat all the way back across the Russian steppes in the merciless Russian winter, which almost destroyed his army.

This triggered an uprising of the eastern European states, spontaneous, save for the fact that it was partly funded by the British. The uprising led directly to Napoleon and his depleted army being defeated by the east Europeans at the Battle of Leipzig in October 1813. Despite Napoleon's army being depleted it was nevertheless a battle with 500,000 on each side. The victorious east Europeans then drove Napoleon back into France. Meanwhile in Spain, Wellington, in May 1813, sallied forth from his winter quarters in Portugal, retook Madrid, routed Jourdan at Vittoria and drove the French from Iberia.

By 1814, after a series of engagements including the Pyrenees, Nivelle, Nive and Orthes, Wellington was in Bordeaux. He then finally brought Soult, Napoleon's remaining leading general, to battle and defeated him at Toulouse. Both Thomas Wildman and Thomas Worsley took part in each of these actions. Napoleon's reign was now in its final days and in April 1814 he abdicated and retired to Elba. However, the leopard never changes its spots and, in 1815, he escaped from Elba and landed in France determined to regain the leadership and, during a triumphal march on Paris, he once more gathered an army about him.

Europe had only just begun to enjoy the sweet taste of peace and to become aware of just how vital it was to give nations that had been at war for twenty years a chance to rebuild. The prize of a share in the potential prosperity of the industrial developments of the nineteenth century beckoned all, but could not be secured without a lasting peace. A metaphorical groan of despair therefore went up all over Europe. Napoleon must be stopped. There were only two nations with armies still in the field, Britain and Prussia, and only one general, Wellington, capable of defeating Napoleon. Wellington, supported by Blücher and his Prussians, met Napoleon in battle at Waterloo in June 1815. In what has been famously and rightly described by Wellington as a 'damn near run thing' Wellington prevailed. This time Napoleon was consigned to the island of St Helena way down the Atlantic under guard which finally brought the Napoleonic Wars to an end and a lasting peace for almost a century.

Both the Light Division and Thomas Worsley were, of course, in the thick of the fighting at Waterloo, the latter rejoining his regiment having been allowed home in recognition of his service and the wounds inflicted on him. Following the battle he was able to report two further strokes of luck. Firstly he was once again struck by a bullet in the neck, and was again thus spared from death by his height but, better still, on the opposite side to the first wound, which had the effect of setting his head straight. Both now returned home, the Light Division to be feted nationally for their remarkable successes and Thomas Worsley on leave to Wallingwells where he was quite rightly received as a hero.

Thomas Wildman was also at the Battle of Waterloo. He had been appointed an extra aide-de-camp to Lord Uxbridge. Henry Paget, Lord Uxbridge, was the highly-regarded commander of the Allied cavalry which numbered some 13,000 during the battle, and Wildman would have been at his side throughout the day. In the early afternoon, at a critical stage in the battle, Lord Uxbridge led an effective charge of the 2,000 strong heavy cavalry against a force of 15,000 French infantry, which had a crucial effect on the eventual outcome of the battle. Unfortunately, he was unable to control the huge momentum of the charge in the recovery stage, not the first cavalryman to have that problem, which enabled the French cavalry to counter-attack the heavies to good effect, but the heavies had already done the damage. Uxbridge spent the remainder of the battle leading several charges by light cavalry units and formations, having eight or nine horses shot from under him.

One of the last cannon shots fired in the battle, and which contained grapeshot, hit Uxbridge's right knee, necessitating the leg's amputation above the wound. It is said he was close to the Duke of Wellington when his leg was hit, and exclaimed, 'By God, sir, I've lost my leg!', to which Wellington replied, 'By God, sir, so you have!' Wildman himself was more fortunate since, although also wounded in the battle, his injury was much less serious. After receiving his wound, Lord Uxbridge was taken to his headquarters in the village of Waterloo, a house owned by a certain Monsieur Paris. There, the remains of his leg were removed by surgeons, without antiseptic or anaesthetics.

Uxbridge remained composed. Thomas Wildman was present throughout the operation and later wrote a letter containing one of the only formally recorded accounts. According to Thomas Wildman, during the amputation, Paget smiled and said, 'I have had a pretty long run. I have been a beau these forty-seven years, and it would not be fair to cut the young men out any longer.' The Prince Regent created Uxbridge Marquess of Anglesey five days after the battle. Lord Uxbridge gave Paris permission to bury the leg in his garden. Later Paris turned the place into a museum. The exhibits included the bloody chair upon which Uxbridge had sat during the amputation. Visitors were then escorted into the garden, where the leg had its own 'tombstone'.

The 3rd Duke of Portland died shortly after his resignation in 1809. On succeeding to the Dukedom, the 4th Duke became politically closely associated with Canning, who was his brother-in-law, and through him accepted Cabinet office as Lord Privy Seal; he was later appointed to the post of Lord President of the Council after which in 1828, he gave up active politics. It was said that 'He possessed great physical energy and mental alertness and his interests, to the last were both keen and numerous'. These became more important to him than a career in politics. When he succeeded to the title he showed himself to be a shrewd manager of his estates, improving them financially. He was particularly interested in farming methods and techniques and undertook several drainage schemes, gaining a reputation as one of the foremost agriculturalists of the day. He was also a keen sailor and amateur yacht designer. Almost certainly he gained this interest in sailing from his time in The Hague, since Holland was the birthplace of sailing as a sport. His interest was in naval design and so went further than mere recreation and continued throughout his life. His most significant contribution was to promote the radical designs of Sir William Symonds, the Surveyor of the Navy, which Symonds created in his spare time. To do this the Duke commissioned Symonds to build a yacht for him during the early 1830s incorporating a wider beam and a more wedge-shaped bottom which, amusingly, he called *Pantaloon*. The Duke then persuaded a reluctant Admiralty to race their best designs against *Pantaloon* which proved to be faster, more stable and to have greater load-carrying capacity.

The 4th Duke was also heavily involved in horseracing, was a tenant of the Jockey Club at Newmarket, and was responsible for many improvements there, including the turf, the gallops and the building of the Portland stand. He won two Group 1 races, the 1819 Derby with *Tiresias*, and the 1838 St James's Palace

Stakes at Royal Ascot with *Boeotian*. His younger brother was Lieutenant Colonel Lord William Charles Augustus Cavendish-Bentinck (1780–1826). Although Lord William never served in the Regiment he did, unknowingly, make two memorable contributions to the history of the Sherwood Rangers: firstly because his great grandson William Henry Cavendish-Bentrick (1893–1977), whilst Marquess of Titchfield, commanded the Sherwood Rangers in the 1930s before succeeding as the 7th Duke of Portland in 1943; and secondly because he was also a great-grandfather of Queen Elizabeth the Queen Mother, who held the appointment of Royal Honorary Colonel of the Royal Yeomanry, of which the Sherwood Rangers Yeomanry formed part, for over thirty years from 1967 till her death.

Chapter 11

A Standing Reserve 1808

The capture of the Danish Fleet in 1808 was not only a tipping point in the Napoleonic War but a crossroads for the yeomanry. During the period between the Regiment being reformed following the resumption of hostilities in 1803 and 1808, when all risk of invasion ended due to the capture, there had been no examples of civil disorder requiring their mobilization. Much work in preparation for the possible invasion, but no civil unrest – clear evidence that the hotheads had, indeed, been kept busy overseas. This was, of course, in contrast to the initial phase of the war when elements of the Regiment had been called out several times in relation to the bread riots. Despite that fact, in 1808 there was no similar order issued to the yeomanries to stand down as had been the case in 1802, although certain troops did take the opportunity to do so unilaterally, in particular the Nottingham Troop for the second time. A surprising decision given that Nottingham was the least stable part of the county but which may be explained by the fact that Lord Middleton and others who would have been funding the Troop, may have felt the cost of doing so could no longer be justified, especially if Ichabod Wright was taking command of the militia. The advantage was that the unit would remain operational to provide local security but at less expense to its supporters.

This is a defining moment in the history of this country's reserves, since it is the first time a formal national infrastructure of reserve units, that is to say the surviving yeomanry regiments, as opposed to random units, was retained in a formed but unmobilized state without the justification of an immediately present threat. It proved to be a good decision.

The Regiment now entered the era of the 4th Duke of Newcastle. On his return from France in 1806 aged twenty-one, having been released by the French, his first priority was to re-establish the close ties that he had with his family. Having done so, he then started to develop an active interest in his inheritance. This included Clumber Park and the surrounding estate and, interestingly, Nottingham Castle, as intrinsically tied up in the legends of the region as was Sherwood Forest itself, part of which he also owned. All of this had, by then, spent a considerable time under the control of his father's executors and his trustees. He also soon became involved with the Regiment. In 1808 the 4th Duke was aged twenty-three, old enough to command a Troop, and showed that he was ready and willing to do so by taking command of the Rufford Troop when Richard Lumley-Savile relinquished command on becoming Earl of Scarbrough. In due course, he renamed it the Clumber Troop.

The extent of Colonel Eyre's involvement at this time is not known for certain. In 1808 he was aged fifty-one. He was still sitting as an MP while his son, Gervase Anthony, was aged seventeen and about to join his father's old regiment, the 1st Foot Guards. As for his daughters, the eldest, as mentioned, had married Charles Pierrepont, heir to Earl Manvers, the second, Frances Julia was destined, in 1814, to marry Granvile Harcourt Vernon, and the youngest, Henrietta, had already married her cousin John Hardolph Eyre, who died in 1817. He was the son of John Eyre, Archdeacon of York. Henrietta later married, secondly, Henry Gally Knight, MP of Langold on the Yorkshire/Nottinghamshire border whose estate bordered that of the Woollaston Whites at Wallingwells and of whom he was a kinsman. This illustrates that Colonel Eyre was still very active and involved in the County.

The role of the Regiment had now changed from the dual one of providing home defence against invasion and as a bulwark against republicanism. Its new role was to maintain law and order against a number of varying pressure points of which republicanism was only one, but an important one and probably the underlying thread between them all.

There is no sign that the move to reorganize the troops into a formal regimental structure in 1805, when invasion was imminent, was implemented. The reason was probably that the threat level fell so quickly thereafter. This meant that, in 1808, the individual troops were no more than affiliated. Some, however, from time to time, worked more closely together, and some not, so the effect was of at least two informal groupings, one in the north and one in the south and the balance who regarded themselves as independent and referred to themselves as such. Undoubtedly Lieutenant Colonel Eyre still commanded the Retford Troop, continued to hold the two standards presented to the Regiment on formation, and enjoyed widespread respect. His role as Commanding Officer if it existed at all which seems doubtful seems to have continued to be one of presiding over a loose confederation of troops which enjoyed a considerable level of autonomy rather than anything more. What seems more likely is that as the longest serving and senior officer amongst the Nottinghamshire troops he would have been invaluable as an adviser to his near neighbour, the 4th Duke who had no formal military training. For convenience the term "Regiment" will continue to be used to describe all Nottinghamshire based yeomanry troops.

The structure of the Regiment in 1808 was as follows:

Troop	Name	Date Gazetted
Retford Troop	Captain Anthony Hardolph Eyre of Grove	8/8/03
	Lieutenant Samuel Crawley	8/8/03
	Lieutenant Stephen Atkinson	24/08/04
	Cornet Samuel Thortel	27/11/04
	Surgeon John Hull	17/12/03
	Ensign Henry Mason	24/08/04
Mansfield Troop	Captain William Boothby of Edwinstowe	8/8/03
	Lieutenant Francis Vickers	8/8/03
	Lieutenant J. Heygate	1/1/10
	Cornet James Maltby	8/8/03
	Cornet G. Walkden	1/1/10
Newark Troop	Captain Francis Chaplin	15/8/94
	Lieutenant Phillip Palmer of East Bridgford	15/8/94
	Lieutenant T. Wright	22/8/08
	Cornet J. Handley	22/8/08
	Surgeon T. Bland	3/4/04
Clumber Troop	Captain the Duke of Newcastle	7/7/08
	Lieutenant R. Nassu Sutton	24/1/17
	Cornet Henry Machin	?
	Cornet R. Milward	25/1/17
Holme Pierrepont Troop	Captain the Hon. Charles Arthur Herbert Pierrepont	
Bunny Cavalry	Captain Joseph Boultby	
	Lieutenant William Timm	
	Cornet Henry Breedon	

However, this table should be qualified by the fact that it is known that the Southern Troops, the Holme Pierrepont Troop and the Bunny Troop, tended to think of themselves as a separate grouping. It is also known that the Newark Troop regarded itself as independent from 1802 until 1828.

The Sherwood Rangers Yeomanry possess three commissions of members of the Machin family into the Regiment, the first of which is that of Henry Machin. At this point Henry Machin was living at Eakring. Henry's father, George, had married Elizabeth Vessey, the sister of the John Vessey mentioned as a Cornet in the Rufford Troop in 1802, who was a tenant of the Lumley-Saviles at North Laithes near Eakring. In 1814 Henry took over the tenancy from John Vessey.

Henry's mother had cousins who owned significant estates at Gateford and Aughton but had no heirs. As a result, when the last surviving cousin died in 1823, he left all those estates to Henry Machin and Henry, now married, duly moved to Gateford and built Gateford Hill which has been the Machins' family home till recently. They still own the surrounding land. The Machins have honoured their good fortune by including Vessey as one of their Christian names ever since.

In 1811 Lieutenant Colonel Eyre's only son and heir was killed in action at the Battle of Barossa in Spain whilst serving with his regiment; he would have felt such a loss keenly. The following year he retired from Parliament and, in 1813, the Retford Troop was disbanded and the Newark Troop was augmented, becoming entitled to two lieutenants, implying that the Retford Troop did not so much as disband as transfer into the Newark Troop. Almost certainly that move would have been triggered by Colonel Eyre's obligatory retirement on attaining the age of fifty-five the previous year. Thereafter, he devoted himself to county affairs, serving as a Justice of the Peace for many years. It is significant that he did not, at that logical stage, hand over the Standards. Clearly he did not consider the new formation to exist as an identifiable unit.

There is a muster roll of the Clumber Troop which has survived dated 1813, which shows the Duke of Newcastle was the Commanding Officer. It also shows that the sixty-five men enlisted out of an establishment of 150, were mostly from Ollerton, but also living at Gamston, South Collingham and Farnsfield, implying that because of the fairly large distances between those locations, about ten miles, that each was a muster point for a detachment. It is consistent with that of the old Rufford Troop plus people from the 4th Duke's own estate.

Also in 1813 the Government, as ever looking for economies, tried to

formalise the informal regimental structure of the Nottinghamshire Yeomanry by amalgamating the Newark, Clumber, Mansfield, Holme Pierrepont and Bunny Troops. However this plan foundered as no funds were forthcoming with which to pay for standardising the uniform.

Although at first sight it seems strange that a regiment which had been raised on a cohesive basis and well led should fragment in this way. Partly the reason was, of course, that, when it was re-raised, the Government had done so, as a matter of policy, on the basis of independent troops. It is also an old truth that in the end the volunteer army is, and remains, a sub-unit based organization, ultimately tending to be more loyal to its home location than to a more widely dispersed regimental affiliation.

There was, however, one way in which the 4th Duke could create an overarching influence over the Nottinghamshire Troops. In 1809, at the age of only twenty-four he had been appointed Lord Lieutenant on the death of the 3rd Duke of Portland. The role meant that all Corps Commanders and Troop Commanders had to clear with him in writing the selection of all officers, who were of course commissioned by him, and report all retirements. They also needed to copy him in on all their strength returns, so that he could monitor their state of operational effectiveness. More importantly, it was, of course, he alone who could order them to 'Assemble for Permanent Duty', so if they were required operationally, only he could issue the order when requested by the Magistrates to do so. That same process had to be gone through to seek his permission to arrange any continuous periods for training; in other words, for a weekend training or annual camp, including the dates and location. Over the years, irrespective of whether he was serving, he took the keenest of interest in these responsibilities, so by virtue of his role as Lord Lieutenant alone, he exerted a strong leadership role which was little short of that of a commanding officer.

This relatively brief period of approximately twenty years in the history of the Lieutenancy was probably the period when that ancient office carried the most responsibility, because never before or since had the Lords Lieutenant had permanently-formed units under their command during a period when they were being called out on a regular basis to maintain law and order. By the mid-1830s the powers of the Lieutenancy were beginning to be eroded.

Chapter 12

The Luddite Riots 1811

Whilst the structure of the Regiment had been evolving it had not been inactive. A period of turbulent unrest began in 1811. In terms of the sequence of events in the war, this was just as, in the Peninsular, Wellington was planning to debouch from behind the lines of Torres Vedras.

The movement was called Luddism, so named after a Leicestershire hothead called Ned Ludd who had destroyed some industrial equipment a generation previously. In general Luddism was defined by opposition to the new machines and mechanized processes of every type which were appearing in the factories. These machines were, of course, leading to the automation of manufacturing processes and this increased productivity while reducing both jobs and wages.

It was an inevitable consequence of the fast developing Industrial Revolution and, whilst it was a national problem, the concentration of the Industrial Revolution in the midlands and the north meant that the worst rioting occurred in these regions. Nottinghamshire was destined to be one of, if not the, epicentre of these riots.

The actions taken by the dissidents became known as the Luddite Riots. The tactics employed by the rioters varied from place to place and from industry to industry. In Nottinghamshire it involved breaking into the factories, mostly textile factories, and destroying either the machines or the frames used in the manufacturing of textiles. Those frames were the key part of the automated stocking-making process which was one of Nottingham's main industries. The task of the Regiment was to stop them and to contain the level of civil disobedience. This contest and related troubles continued on and off for seven long and violent years, a period which did not end until well after the Napoleonic Wars had finally been concluded at Waterloo,

The first outbreak of trouble took place on 11 March 1811. Several hundred people assembled in the Nottingham Market Place. As night fell they marched to Arnold on the outskirts of the town and broke sixty-three frames. It is likely to have been the 4th Duke's first challenge whilst in command of a Troop and indeed whilst Lord Lieutenant. In his capacity as Lord Lieutenant, the 4th

Duke had presumably reacted to the disturbance in Arnold by calling out the Clumber Troop, now commanded by himself, and the Holme Pierrepont Troop commanded by the Hon. Charles Pierrepont, representing between them two of the four Dukeries families. The rioters had dispersed by the time they got there.

It is interesting to reflect on the logistics of calling out troops of yeomanry. First of all the news of the riot would need to get through to the authority locally, presumably in this case the Chief Constable on application to the Magistrates, who would then take time to assess the situation before deciding that it was sufficiently serious to justify a request to the Lord Lieutenant to call out elements of the yeomanry. That request would then need to be communicated to the Lord Lieutenant some twenty miles away at Clumber by a messenger on horseback. The Lord Lieutenant would certainly need Lieutenant Colonel Eyre to report to him so they could decide how to play it. It is hard to imagine the process so far taking less than a day but probably not more than two.

The next stage would be to get the word to the two Troops whose members lived up to ten miles from where their Troop mustered. One imagines they would have used some form of cascade system of notification on horseback, but even so it would not be something which could be achieved in an instant. Indeed this part of the process was described as 'assembling', a word which indicates that it was something that took some little time. By the time the Troop members had made arrangements for their absence from their farms and businesses and 'assembled', it is hard to imagine much less than another day would have passed. They would also need to make suitable logistic arrangements, for example the supply of fodder.

They then had to make the twenty-mile march to Arnold at a pace, usually a steady trot, interspersed with a period of walking achieving an average of about six miles per hour that would result in them having some horse left under them when they got there. On that basis the whole process would take at least three and probably four days, not much less than the five days in which the Sherwood Rangers Yeomanry, when part of a NATO-assigned armoured reconnaissance regiment during the Cold War, were required to be able to reinforce BAOR in West Germany, so it is hardly surprising that the rioting was over by the time they got there. It was also clear that the process would need streamlining. The two Troops remained under arms for some days, patrolling through and billeted in the local villages, and were then dismissed.

On 10 November there was fresh trouble, this time in Bulwell, where armed rioters attacked the house belonging to one Hollongsworth 'who had become obnoxious' because he was using the new sort of frames in his business. He had become aware of the rioters' intentions and, assisted by some friends, barricaded himself into his house, where he had taken the frames for safety. Shots were exchanged and one of the rioters was mortally wounded. The rioters were then reinforced by others and their mood was inflamed further as their wounded comrade died in the road before them. They then renewed their attack with such vigour that Hollongsworth and his colleagues were forced to withdraw, leaving the house which the rioters then entered and smashed not only the frames, but everything in it.

Many owners were seeking to protect their frames by transporting them to Nottingham where they could be protected centrally. The next day the mob intercepted some frames in Basford being moved in two wagons to Nottingham for this purpose, demolished them on the road and threw them in the River Leen. On 13 November the rioters turned their attention to Arnold where they destroyed thirty-seven frames.

The Bays, the regular regiment deployed to help deal with these problems, had become so stretched that they were able to be of little effect and the 4th Duke therefore, in his capacity as Lord Lieutenant, now called out the whole Regiment, then consisting of the Newark, Clumber, Mansfield, Holme Pierrepont and Bunny Troops. The Duke, having learned lessons from the earlier occasion, had given orders for them to assemble some days before and already had them stood to nearby. This enabled the Holme Pierrepont and Bunny Troops and a detachment of the 2nd Dragoon Guards to react immediately to check the disorder. They came on the mob at Sutton and instantly dispersed them.

Henceforth it increasingly became common practice to seek to overcome the length of time it took Troops to assemble by anticipating the likelihood of them being needed and calling them out early enough to be ready when trouble broke out.

A large number of prisoners had been secured at Sutton and taken to Mansfield, and so over the next two days, 14 and 15 November, were moved in post chaises to Nottingham gaol escorted by elements of the Regiment. The Regiment now patrolled the villages where the troubles had flared up and the surrounding areas for several days. There was fresh trouble at Old Radford on the 18th where some frames were broken. The Regiment hurried to the village and the mob fled.

The Newark Troop, based on Southwell, was only out for two days. The Bunny Troop was out for six days and the other three troops were out for nine days before being allowed to go home.

The first week in December saw trouble afoot again. The Bunny Troop marched up to Sutton-in-Ashfield where there had been an outbreak, and the Mansfield and Holme Pierrepont Troops also assembled. These Troops were stood to for about a fortnight, which initially achieved stability, but now the troublemakers changed their tactics and resorted to night raids and the frame smashing was spreading to the neighbouring counties. The Holme Pierrepont Troop was stood down on 12 December, having been on duty for seventeen days in the previous two months and the Mansfield Troop was stood down on 19 December.

By January 1812, as Wellington was preparing to march on Cuidad Rodrigo and Badajoz with young Thomas Worsley in the line, there were three regiments of militia from outside the county, as well as the Bays, all stationed in Nottinghamshire. In addition, Bow Street officers were being sent from London and the yeomanry regiments in all the surrounding counties were stood to in their own counties. The high profile military presence conveyed by constant patrolling and picqueting gave the whole region a warlike feel. Notwithstanding all these steps the rioting recurred in Nottingham and the surrounding counties on a daily basis, and had become more complex because the rioters had become increasingly adept at carrying out their raids despite the heavy patrols, by assembling and dispersing quickly, destroying over 900 frames in the process.

The following extract is from Benson Freeman's excellent history of the South Notts Hussars which describes the Regiment's involvement:

On January 13th 1812, a riot took place at New Radford, and at the same time twenty frames were destroyed during the night at Lenton within a few hundred yards of the barracks. After doing this the Luddites crossed the River Trent and broke fourteen frames at Ruddington, and twenty at Clifton, leaving only two whole frames in the latter townships. An express was sent to Nottingham for a troop of Hussars, who went off with all possible speed with as many of the Bunny Troop as could be collected. Those of the Bunny Troop who were in the immediate neighbourhood of the scene of action were immediately mounted and galloped off, one party to pursue the depredators while other detachments of the Troop rode 'desperately' and secured all the passes over the Trent for a space of four miles, under the expectation of intercepting the rioters on their return to the town. The Luddites, however, seized a boat above Clifton, and on arrival at the opposite bank discharged their firearms and made good their retreat, breaking off into two divisions.

Although the meeting of armed mobs was rendered impossible by the large force of military in the area, there were still cases of frame breaking throughout the year. At Basford, one case was most daring. A party of three soldiers were protecting a house; a party of Luddites entered the house, and immediately confined the soldiers, and while two of the party stood sentry at the door with the soldiers' muskets, the other demolished the frames. After the mischief had been done the muskets were discharged and the soldiers liberated, the depredators wishing them good-night. At Nottingham the rioters gave up frame breaking for worse atrocities, and even placarding the streets one night offering a reward for the Mayor 'dead or alive' because he had offered a reward of £500 from the Corporation of Nottingham for this discovery of the assassin who shot Mr Trentham. But by the beginning of

April, thanks to the Military and Yeomanry, the more violent part of the business was over as far as the County was concerned.

Regardless of the fact that the Yeomanry were the utmost assistance to the Civil Power during these disturbances, they do not seem to have been over well treated by His Majesty's Ministers, for on April 27th, 1812, Lord Sidmouth writes to the Lord Lieutenant of Nottinghamshire regretting that he cannot allow any grant to meet the 'additional expenses which have been incurred by the individuals of the Bunny and Holme Pierrepont Troops of Yeomanry Cavalry during the period of their being assembled at Nottingham for the suppression of disturbances'.

Matters eventually abated in April but a swathe of damage was left through much of Nottingham.

The 1st Earl Manvers died in 1816 aged seventy-eight, having lived long enough to celebrate the victory at Waterloo. He had been a great supporter of and mentor to the Regiment, having willingly stepped into the vacuum left by the unexpected death of the 3rd Duke of Newcastle in 1795 and had been at the heart of the Regiment's development in the twenty years since, and one likes to think that this contribution was a significant factor in his various ennoblements. His sons had also raised and commanded more than their fair share of Troops throughout the period. He was succeeded by his third, but oldest surviving, son, Charles Arthur Herbert Pierrepont (1778-1860), as the 2nd Earl Manvers. He was aged thirty-eight and, as a long time Troop Commander, was also a strong supporter of the Regiment.

Chapter 13

The Derbyshire Rising 1817

The post-Waterloo era had now dawned. The country was at last at peace with its neighbours if not with itself. It is ironic that this had been a war fought ostensibly to defeat once and for all a combination of republicanism and a challenge to the established order, namely, rule by an unelected elite in whom, alone, real power resided.

However it had failed to have that effect because, when the dust of Waterloo had settled, it was clear that those same problems, far from being defeated at Waterloo, had been elevated by the war from being merely discernible as issues before, to being widely supported and irresistible popular movements afterwards. History indicates that the same effect can be seen in the case of most major wars, which poses the question of whether such wars, though they seem to be primary events at the time, merely serve to delay the birth of the wider change which they fail to suppress, and in the end nothing can resist.

Although republicanism was still alive as a movement, the most prominent matter, which was now much more clearly defined, was a movement leading towards the vote for all, which became known as Reform. However, the Tories, effectively in power since 1783, having thought they were fighting Napoleon for the status quo and having won a great victory at Waterloo, were in no mood for change. They could not see the inevitability of the changes and that the war had, in effect, stopped nothing except Napoleon himself. In this misguided stance they were supported by the Whigs, who were the main opposition party. Having misread the situation neither could, at that stage, grasp that what was coming would, in the end, prove beyond their power to stop.

Lord Liverpool presided over the Government with 'tact, patience and laxity'. His leading ministers were Lord Castlereagh, his status greatly enhanced by the skill with which he had negotiated the peace settlement. In addition, from 1818 onwards, Liverpool was joined by the Duke of Wellington, intent on resuming his political career. Both Government and Opposition seemed incapable of distinguishing the demand for Reform, which had been strengthened by the accelerating pace of the Industrial Revolution, from the practical effects of the Industrial Revolution itself, namely inner-city squalor, poor housing, poor pay, and high food prices, all exacerbated by a post-war recession.

If the politicians had just sought to address the latter demands, which were each capable of being dealt with, and should have been, they might have been able to fend off the matter of Reform for longer, but their only response was to lump all these problems together, misidentify them as an attack on property, and set about defending property, largely owned by the ruling aristocracy, from that attack.

They were not imagining the attacks which were real enough and were coming from a new, and as yet minority, group who described themselves as Radicals. Their beliefs made them effectively, the only opposition, but they were, of course, unable to promote their views on the floor of the House of Commons because they were unable to get elected. Instead they forfeited the moral high ground by resorting to civil disobedience, drawing their support from the general discontent felt amongst the working classes in the northern counties. Their actions were initially counter-productive because they frightened those elected opponents of the Government, who might otherwise have been persuaded to support them, into siding with the Government. The troubles were initiated by agitation by ringleaders, consisting of inflammatory speeches and quasi-military-type activities, aimed at actual uprising and civil war. The stakes were high.

The most high profile protest was that of the Blanketeers who, in 1817, marched from Manchester to London. They were so called because they carried blankets for sleeping in on the way. Their march was easily broken up. In the same year the next most prominent event which took place was called the Derbyshire Rising.

The threat of insurrection in Nottingham was perceived to be so great that in February 1817 it had been resolved to hold the Assizes at Newark rather than Nottingham, but the Prince Regent, the future George IV – George III now being too indisposed to govern due to mental illness – refused to ratify the decision, so the assizes went ahead as planned, but on the back of extensive measures to make sure that any trouble could be contained, including the warning of the Regiment for service.

The ringleader of the Derbyshire Rising was a man called Jeremiah Brandreth, and the rising was therefore also referred to as the Brandreth Riots. Jeremiah Brandreth's cause was republicanism. The troubles themselves started around March 1817 in Derbyshire with Brandreth making many violent speeches and, furthermore, manufacturing large quantities of pikes and purchasing firearms. His purpose was nothing less than civil war. In a related and simultaneous incident there were also riots in Ruddington causing the Bunny Troop to be called out for two days to restore order there.

On 11 April 1817 the Secretary of State for Home Affairs, Viscount Sidmouth wrote to the 4th Duke

Vigilance and Precaution is still necessary [and to keep the] Watch and Ward Act in operation for a month or six weeks longer in the manufacturing districts of Nottinghamshire.

The situation in neighbouring Derbyshire was much worse however. During the first week of June 1817 increased activity amongst Brandreth's followers was noted, and it was clear that he was about to make his move. Encouraged by a government spy called Oliver, Brandreth then set off from Pentridge and Ripley in Derbyshire leading his supporters, who were mostly made up of some 300 poor stocking-makers armed with pikes, pistols and firearms. They were promised that they would receive 100 guineas, bread, meat and ale. The plan was that they would then lead an attack on the local barracks and overthrow the government. It was clear that they were headed towards Nottingham to join their brother insurgents, for whom they had developed great admiration because of their achievements during the recent Luddite Riots. Brandreth had told his followers that the whole country was about to rise up with them and that a provisional government was to be set up to provide relief from their grievances, the nature of which have been referred to earlier, and 'end poverty forever'.

Naturally this development was noted with some considerable concern by the authorities. On Sunday 8 June the Nottingham Magistrates established themselves at the Police Office determined at all costs to prevent the insurgents from joining up with their Nottingham sympathizers. The magistrates stayed there throughout Sunday and Sunday night and Monday and Monday night, and monitored the situation closely. Having warned the Mansfield, Newark, Holme Pierrepont and Bunny Troops for action, the Lord Lieutenant, was also prepared. They were now stood to together with the 15th Hussars and the 95th Regiment of Foot.

At first light on the Tuesday morning Oliver alerted the authorities that the insurgents had reached South Wingfield in Derbyshire, confirmed they were armed, and were heading with increasing numbers for Nottingham. He also reported that one man had been shot dead in his house for not yielding up arms and that 'acts of outrageous violence were perpetrating'. Orders were immediately issued to the Newark and the Holme Pierrepont Troops to march their Troops at once with all despatch to Nottingham, part of an extensive plan which also included the simultaneous deployment of the 95th of Foot, and the raising of special constables. The Mansfield Troop was stood to in Mansfield and orders were sent to warn the Bunny Troop.

Two magistrates, Lancelot Rolleston, who would raise the Watnall Troop immediately following this incident, and C. G. Mundy, were sent with elements of the 15th Hussars to intercept the rioters. They came up with the advance guard at Gilt Brook near Hill Top, which immediately broke and ran. They were promptly pursued from the area of Kimberley through Eastwood back into Derbyshire towards Langley Mill and about thirty of them were captured along the route, many with arms in their hands. The Hussars had been assisted especially in rounding up the actual ringleaders by the Chesterfield Troop of the Derbyshire Yeomanry, who arrived at a timely moment in the rear of the insurgents, and returned with their prisoners and a considerable haul of weapons of various sorts at about 6 o'clock in the evening to secure them in the county gaol.

This timely intervention brought that particular incident to an end, and the Regiment was stood down after three days. The Nottingham Journal was full of praise for all the Troops involved and the alacrity with which they had responded to the emergency. Brandreth himself was captured a few days later, eventually tried for high treason and hanged in Derby in November 1817. This brought only temporary respite and general unrest continued.

The whole country was by now on the verge of open rebellion over republicanism, intertwined with Reform, which had to be contained. This period brought significant concessions from the Government towards service in the yeomanry which yielded a strong response from the gentry and the middle classes alike who were all so deeply concerned at the seriousness of the threat. In Nottinghamshire an enlargement in the number of troops was triggered, of a scale never seen before or since. It was felt that only a major show of force and readiness to be able to respond to any hint of trouble would pre-empt any serious uprising.

In 1817 the 4th Duke, showing a lead, raised a second Clumber Troop. In addition two new Troops were raised that June, in or around Nottingham, one, as mentioned, at Watnall by Lancelot Rolleston and another at Wollaton, by Lord Middleton who had previously raised a militia.

Records show that in 1817 the Southern Troops were officered and manned as follows:

Troop	Officers	Strength
Holme Pierrepont Troop	Captain Jonas Bettison	88 all ranks
	Lieutenant William Charlton	
	Cornet Gilbert Maltby	
Bunny Troop	Captain J. Boultbee	41 all ranks
Watnall Troop	Captain Lancelot Rolleston	88 all ranks
	Lieutenant Thomas Barber	
	Lieutenant Samuel Potter	
	Cornet George Robinson	
Wollaton Troop	Captain Henry Willoughby	64 all ranks
	Lieutenant Thomas Webb Edge	
	Cornet W. Pepper	
	Surgeon William Wright	

Captain Henry Willoughby (1780–1849) was a kinsman of the 6th Baron Middleton and is stated in the Middleton family tree as being of Apsley Hall in Warwickshire and Birdsall in North Yorkshire, both Middleton estates. He was married to Charlotte Eyre, the daughter of the Venerable John Eyre, Anthony Haraolph Eyre's brother. Although not a close relative of the 6th Baron, he was effectively the 7th Baron's heir so when the 7th Baron died, after Henry Willoughby's death, the title passed to Henry Willoughby's son and has continued down that line since.

Captain Boultbee died in 1818, and the Bunny Troop disbanded itself. The reason was that he had raised it and been its driving force and no successor could be found. About the same time, the Holme Pierrepont Troop raised a second troop which was called the Bingham Troop, the 2nd Earl Manvers wasting no time in demonstrating his own support for the Regiment. In June 1818 the Watnall and Wollaton Troops were called out to assist in the suppression of further republican riots, providing yet another spur for further expansion.

In 1819 several other troops were raised. The Nottingham Troop was re-raised by B. Freeman and took upon itself the name of the Loyal Nottingham Troop of Yeomanry Cavalry. They claimed to be an independent troop. Given the many disturbances in and around Nottingham during the ten years since the Nottingham Troop had stood down, this move was not before time. There were now no fewer than five Troops either in or surrounding the immediate outskirts of Nottingham, namely Nottingham, Holme Pierrepont, Bingham, Watnall and Wollaton, effectively an entire regiment's worth, as clear an indication of the degree of civil unrest in the town as could be given. It also gave those five troops a similar focus which they shared in common, but which did not concern the northern troops of the Regiment in the same way.

However the rest of the county had concerns aplenty of their own judging from the expansion there.

At the outset of 1819 there were five troops in the north, two based on Newark, two based on Clumber and one on Mansfield. By the end of the year the position was as follows:

Newark had reduced to one troop.

Clumber had maintained two troops and was now a Corps. The 4th Duke, as its commander, was promoted to the rank of Major.

Mansfield continued unchanged save only that Major Boothby, who had commanded since 1794 and had been second in command of the original regiment, retired.

Retford raised two troops, confirming that Newark had in effect been looking after the rump of the old Retford Troop.

Worksop reformed.

This made a total of seven troops in the north and a total of twelve for the county, representing an establishment of approaching 1,000 all ranks for Nottinghamshire as a whole if fully recruited.

1819 was the year of the 'Peterloo Massacre', the emotive name given to a clash between the rioters in Manchester, whose demand was Reform, and the Manchester and Cheshire Yeomanries. This became symbolic of an over reaction by the forces of law and order, but which the more objective historians assess as having been less bloody than incidents in some other locations including Nottingham. However, self evidently, the Regiment can take great credit for the fact that, despite the more volatile climate in Nottinghamshire, they discharged their role without ever being criticized, or accused of overreaction, or of losing control of the situation.

Later the Duke of Wellington, whilst giving evidence before the Parliamentary Finance Committee said,

It is much more desirable to employ cavalry for the purpose of police than infantry; For this reason, cavalry inspires more terror at the same time that it does much less mischief. A body of twenty or thirty horses will disperse a mob with the utmost facility whereas 400 or 500 infantry will not effect the same object without the use of their firearms and a great deal of mischief may be done.

In Nottinghamshire, this period included one precautionary call out of all the southern troops in Nottingham, which included several other units and significant logistical arrangements over several weeks, but the call out was later found to have been based on false intelligence.

Chapter 14

The Northern Troops 1819

There was now a discernible divergence between the objectives and role of the southern troops, and those troops, for now referred to as the northern troops. Since the northern troops would, within the next decade, form a single regiment which was to be called the Sherwood Rangers Yeomanry, a more detailed description of them and the people who served in them is relevant. The 4th Duke went one step further than had the southern Troops. He decided, in 1819, to form as many of the northern Troops as were willing, formally into a new regiment which he named the Clumber Yeomanry. The Troops and the officers were as follows:

Troop or Corps	Name	Date Gazetted
Newark Troop	Captain Francis Chaplin	21/11/94
	Lieutenant P. Packner	?
	Cornet J. Handley	22/8/08
Clumber Corps of Yeomanry	Major the Duke of Newcastle of Clumber	10/11/19
1st Troop	Captain The Marquess of Titchfield of Welbeck	?
	Lieutenant J. E. Denison of Ossington	10/11/19
	Cornet P. R. Falkner	3/11/19
2nd Troop	Captain Henry Simpson of Babworth	10/11/19
	Lieutenant W. H. Barrow	1/4/20
	Cornet S. E. Bristowe	02/21
	Surgeon Isaac Timms	08/20

Troop or Corps	Name	Date Gazetted
Mansfield Troop	Captain Lord Frederick Bentinck	10/11/19
	Lieutenant J. Heygate	1/1/10
	Cornet J. Rolfe	16/4/23
Worksop Troop	Captain The Earl of Surrey of Worksop Manor	24/6/19
	Lieutenant The Hon. George Monckton–Arundell of Serlby	?
	Lieutenant Sir Thomas Wollaston White Bt. of Wallingwells	5/6/24
Retford Corps of Yeomanry and Riflemen		
1st Troop	Captain J. Kirk	17/12/19
	Lieutenant Thomas Taylor Worsley	17/12/19
2nd Troop	Captain H. Hutchinson	17/12/19
	Lieutenant R. Corring	17/12/19

The extent to which the troops trained together as a single unit is not clear. Through his role as Lord Lieutenant, the 4th Duke would have had significant influence because he would have had the duty to authorize all training. However that would not be conclusive: the Newark Troop, for example, still under command of the remarkable Captain Chaplain, as they had been since all but the first three months of their being raised in 1794, were still as independent as they had been since 1802 when they "stood alone". There are no records of the Retford Corps having trained with any of the other troops, but there is evidence that the remainder, that is to say the Clumber Corps, the Worksop Troop and the Mansfield Troop, did train together, and were more closely affiliated. Indeed there is evidence that one of the two troops forming part of the Clumber Corps, the one listed second, called itself the Retford Troop and the first called itself the Clumber Troop.

Many of the estates involved in raising troops during this expansion had raised militia in previous expansions. The option of raising a militia was not available, of course, because the Government had disbanded the militia in 1816 in response no doubt to the ending of hostilities. The reasons for having concerns about maintaining the militia but not about raising troops of yeomanry no doubt remained the same as they had at the beginning of the war.

Nothing reflects the severity of the threat at this time more than the senior names now collaborating in the command of these northern Troops. Clearly they must have been deeply concerned, possibly more deeply concerned in relation to the threat of civil unrest, than at any time since the French Revolution. They represent a remarkable group of people:

The Marquess of Titchfield (1796–1824) was William Henry Cavendish Cavendish-Scott-Bentinck, the eldest son and heir of William Henry Cavendish-Bentinck, 4th Duke of Portland (1768–1854) who was the eldest son of the 3rd Duke and had succeeded to the title on his father's death in 1809. The Marquess graduated from Oxford with a Classics Degree, and entered Parliament in 1819 as a Tory MP, initially for Bletchingley and then for King's Lynn. Grenville said of him that he was 'much the cleverest member of the family'. Though also being described as 'eccentric and indolent' he was stated to be able to 'make himself master of any subject he thought fit to grapple with'. Initially, he was thought to be an admirer of Canning but later tended to distance himself from all political parties. He specialized, with some distinction, in finance and the economy. Sadly the Marquess died aged only twenty-eight in 1824. It was his younger brother who eventually succeeded to the title when the 4th Duke died in 1854. As will be mentioned his untimely death may well have deprived the Regiment of a natural successor to the 4th Duke of Newcastle as commanding officer. Not only was he the right age and had appropriate experience, but the Dukes of Portland and Newcastle were quite closely allied politically, particularly in relation to Reform. It was a tragic loss.

Captain Lord Frederick Cavendish-Bentinck was the fourth and youngest son of the 3rd Duke of Portland and uncle of the Marquess. He was thirty-eight years old having had a career in the regular Army during which he had commanded the 58th of Foot and risen to the rank of major general. He was destined to be the great-grandfather of the 8th and 9th Dukes of Portland. His involvement with the Mansfield Troop clearly indicates the family's concern to look after and protect their significant interests in Mansfield.

The Earl of Surrey, Henry Charles Howard (1791–1856). The 11th Duke of Norfolk had died in 1815. Despite having been married twice he had no heir. Unluckily for him his first wife died in childbirth within a year of their marriage and his second became insane within a similar timescale. Although his marriages were childless he later had several mistresses, one of whom bore him five children. For this reason his legal heir was not in direct line but was his third cousin Bernard Edward Howard, 12th Duke of Norfolk (1765–1842) whose father Henry Howard of Glossop in Derbyshire, had been a financially unsuccessful wine merchant in Dublin in the eighteenth century and had been bailed out by the 11th Duke.

The Earl of Surrey was the 12th Duke's only son. In 1827 he became a Privy Councillor and, in 1829, a Liberal MP for Horsham and then for West Sussex, and, therefore, in favour of Reform. He was, of course, a Catholic and his election followed immediately on the passing of Roman Catholic Emancipation which had survived the opposition to it of both the 4th Duke of Newcastle and the 2nd Earl Manvers. He was, therefore, the first person, confessedly a Catholic, to sit in the House of Commons since 1688 but in 1851 he became an Anglican. He also changed his political affiliations and became a Liberal and therefore in political opposition to the Dukes of Newcastle and Portland. At about the time he was in the Regiment, Thomas Grenville described him as 'plain, unaffected, reasonable and good-natured'. He was, by contrast, mischievously described by the 4th Duke of Newcastle in the privacy of his diary as 'miserable rat-like looking', not a polite way to describe a brother officer. He succeeded to the Dukedom as the 13th Duke in 1842 and held all manner of appointments including Knight of the Garter, Treasurer of the Household, Master of the Horse and Captain of the Yeoman of the Guard.

Lieutenant Denison: He was John Evelyn Denison (1800–1873) son of John Denison of Ossington, one of the founding officers of the Regiment. He was a Liberal MP for a variety of seats, including several in Nottinghamshire from 1823 to 1872, a period of forty-nine years. In 1857 he was chosen unanimously to be Speaker of the House of Commons, an appointment he held for fifteen years. On relinquishing the office of Speaker in 1872 he was appointed Viscount Ossington of Ossington 'in consideration of the dignity, knowledge, and ability' with which he had discharged that office. He died only one year later without a direct heir, so the title did not survive. In 1827 he had married Charlotte, the daughter of the 4th Duke of Portland and the sister of the Marquess of Titchfield and a formidable person in her own right.

Specifically remembered for creating the convention that the speaker should use his casting vote in support of the Government. These are a selection of the momentous events over which he presided during his period as Speaker: at home, during Palmerston's premiership, 1855–1865, Free Trade, the 1867 Reform Bill, which redistributed seats in favour of the industrial cities and away from the rural areas (as will be seen, one of the objectives of the Chartists), the premiership of Gladstone, 1868–1874, described as 'one of the best instruments of government that ever was constructed' and which introduced reform across the whole field of public services; abroad: the Indian Mutiny, the Expansion of British Imperial rule worldwide, the American Civil War and slavery, the rise of Germany and the coming to power of Bismarck.

Lieutenant Monckton: He was George Edward Arundell Monckton-Arundell (1805–1876). He became the 6th Viscount Galway in 1834 and was the father of the Viscount Galway destined to command the Regiment and a Conservative MP for East Retford from 1847 to 1876.

Sir Thomas White: He was the second Baronet as the first Baronet, also Thomas, had died in 1817 aged fifty-one. The young Sir Thomas had, in 1820, aged nineteen, taken a commission by purchase in the 16th Regiment of Light Dragoons but had transferred in 1822 from the 16th Light Dragoons to the 3rd Regiment of Light Dragoons by swapping his commission. This move was on the insistence of his guardians, Sir Frederick Gustavus Fowke Bt. and Henry Gally Knight MP, who was married to Henrietta Eyre, Colonel Eyre's youngest daughter. They, contrary to his own wishes, would not permit him to go to India with the 16th Light Dragoons.

Whilst he was serving in the 3rd, the Commander in Chief, the Duke of York, came to inspect the regiment one Sunday at Brighton. The usual practice is to pass behind the officers to inspect the men. Instead, on this occasion, the Duke walked down the front of the line of officers and stopped in front of Sir Thomas. He said 'Your name is Sir Thomas Wollaston White?' to which Sir Thomas answered in the affirmative. 'Your father raised, armed and clothed a regiment of volunteers at his own expense?' Sir Thomas replied that he had. 'A very noble deed, a very noble deed. Attend my levee, Sir.'

For some reason, maybe because he knew what would be proposed, Sir Thomas did not do so, and some time later his commanding officer received a letter from Sir Herbert Taylor, the Duke's military secretary, desiring Sir Thomas' attendance. On his appearance Sir Herbert enquired why Sir Thomas had neglected His Royal Highness' command and saying he was authorized to offer Sir Thomas a lieutenancy without purchase, a valuable offer to a cornet. However, this would have meant leaving the 3rd Regiment of Light Dragoons because no such vacancies existed in it. Sir Thomas declined because he wished to stay with his friends. The offer was left on the table however, but was never taken up.

Sir Thomas left the army in 1824 in order to marry Georgina Ramsay the youngest daughter of George Ramsay of Barton and Sauchie. She was eighteen. Sadly she died only a year later, one suspects in childbirth, leaving a daughter, also Georgina.

It is not necessary to look any further than Lieutenant *Thomas Taylor Worsley* to find the explanation for the reason why the newly-raised Retford Corps of Yeomanry and Riflemen was called thus. The local hero of the Peninsular Campaign had, in 1818, transferred into the 45th Regiment of Foot, the Nottinghamshire Regiment later, in 1881, named the Sherwood Foresters, on half-pay and therefore would have come home. The Sherwood

Foresters and the Sherwood Rangers are often confused. The former were, of course, a very famous regular infantry regiment with a fine fighting record, raised in Nottinghamshire and Derbyshire and having no connection with the Sherwood Rangers.

Thomas Worsley would have been the last person to stand by whilst a challenge to all he had fought to preserve was mounted by Radicals, so it is no surprise to find him in this new role. The fact that there were clearly other Riflemen living in the area with whom he combined to join the new Corps, suggests that the first Sir Thomas had indeed directed volunteers to the Light Division, and it would also be no surprise to find that a number of those recruited to the new Corps had served in Sir Thomas' 3rd Nottinghamshire Regiment of Local Militia, which would have been disbanded only two years previously. The Woollaston White family were very proud of all Thomas Worsley had achieved and this was carried through to his death in 1851 when a plaque to his memory and in particular commemorating his achievements in the Napoleonic Wars was erected in the Woodsetts Parish Church, just over the border from Wallingwells in Yorkshire,by the second Baronet where it can be seen to this day.

Captain J. Kirk had served in the 24th Light Dragoons.

The 4th Duke of Newcastle is left to last because he was much the most influential and involved. By 1819, when he was thirty-four years old, he had accumulated a number of appointments in addition to those already mentioned. He had been appointed Steward of Sherwood Forest in 1812, as the 2nd Duke had been before him, and invested as a Knight of the Garter the same year, all of which, considered alongside his appointment as Lord Lieutenant, indicates the degree to which the family's formidable influence had survived into another generation. His memory is perpetuated at Eton by the Newcastle Scholarship, which he founded,

He was heavily involved in politics, not directly as the 1st Duke had been, but by exerting influence as his grandfather, the 2nd Duke, had done. His political activities were guided by his staunch religious and high Tory principles. He was a supporter of the traditional establishment of church, country and state, and deeply concerned with all the other major issues of his day, and corresponded widely with those in public office.

An influential electioneer, he was regarded as a man worthy of some consideration by the leading politicians of his day, despite the fact that he played no direct role in politics himself. He returned four nomination MPs and was able to influence the return of three others, though his political influence was considered by contemporaries to be much wider – as many as twelve to fourteen nominees was a commonly accepted figure.

His main areas of patronage were in elections in Nottinghamshire, Yorkshire and Falkirk in Scotland. His greatest powerbase lay in the boroughs of Aldborough and Boroughbridge in Yorkshire while his power in Nottinghamshire was somewhat weaker, though he was certainly very influential in Newark where his candidates were generally returned unopposed. He was also an influential figure in local politics and took an active part in Retford, Bassetlaw and Nottinghamshire elections.

It was however the way he exerted his influence which caused censure. He used a variety of means, the incentive of low rents, an extensive network of influential agents, clever use of purchasing power for provisions and goods at Clumber and, even, the use of bribery and treating. These activities made him a highly controversial figure and, as would be expected of one who wielded such power, he had at least as many powerful detractors as supporters, being described by the Earl of Selborne, one of the former, thus: 'He was not a wise man; there was nothing either in his very aristocratic politics, or in his strong political Protestantism to help to form the character of his son.' The *Dictionary of National Biographies* states that 'For more than twenty years the general public censured the Duke's motives as a landlord and a member of the House of Lords and his appetite for jobbery was declared to be insatiable.'

These quotes give a strong indication of how he was viewed locally, which reinforces the assumptions made earlier as to why the Regiment was not more cohesive under his leadership, and also why the southern Troops were keen to distance themselves from the northern Troops which the 4th Duke led.

On the basis of the foregoing it is unsurprising to find him a vehement opponent of Reform. He was after all one of the key people in the country whose power the Reform movement was intended to curb.

To summarize; the northern Troops consisted of twenty officers amongst who were: three descendants of two different Prime Ministers; one Knight of the Garter; one future Knight of the Garter; one Duke; two heirs to Dukedoms; two future viscounts; one of the most influential political operators in the country; a future Earl Marshal of England; a future Speaker of the House of Commons; the first Roman Catholic to be elected to the House of Commons in over a century; four Members of Parliament; and last, but not least, a true hero of the Peninsular Campaign. It is unlikely to have been friendship that brought them together at this time, but necessity.

George III died in 1820 and was succeeded by his son George IV.

The Clumber Yeomanry camped together for the first time on 2 May 1820 for about a week. It was said that because of the high standing of 'some of the officers that many men of great respectability went into the ranks and were mounted and turned out as well as their officers'. Although the period of tension continued in Nottinghamshire until 1824, and, as a result the northern Troops were kept at a high state of readiness, there were in fact no further disturbances which involved the Regiment. Twelve troops of yeomanry clearly acted as a significant deterrent initially. Then an improving economy created a political and civil climate which was much more settled than for many years.

The Regimental records indicate that the 4th Duke commanded until 1828, but it is now clear from his papers that he in fact relinquished command in 1824. This is confirmed also by his diaries which contain an entry for 15 February 1824 which reads 'I have too many irons in the fire – I intend to resign from the Command of the Yeomanry'. The discrepancy between the two dates almost certainly arises from confusing his role as commanding officer with that of Lord Lieutenant which he continued to perform in the same hands-on fashion as he had hitherto. By 1824 he had been in command for seven turbulent years, whether of the wider Regiment or of the Clumber Yeomanry. In addition his wife, Georgiana, had died in 1822, a loss he felt deeply and which left him with twelve children to bring up on his own, a responsibility he took very seriously and which, together with his property interests and his involvement in politics, thenceforth took up all his time. She was the daughter of the widow of the 4th Baron Middleton of Wollaton. Lord Middleton had died in 1781 and his widow had remarried one Edward Miller Mundy of Shipley and, therefore, was related to Mrs Lumley-Savile.

The 4th Duke was entitled to look back on the period with some pride because it had been a period when events had tested the metal of the wider Regiment, which was manned with people who were not independent of the issues, indeed had the most to lose from the changes being campaigned for, but had nevertheless performed with inpartiality, courage, commitment and restraint, and without loss of life in situations when, on any number of occasions, one unwise move could have escalated into bloodshed as it had elsewhere. His decision to retire and confine his involvement to that of his role as Lord Lieutenant was probably a good thing, because he was undoubtedly a controversial, and often tactless, figure who had ruffled feathers. What was needed now was a leader, who could exercise a unifying influence in respect of the wider Regiment to meet new demands, and this he neither could, nor did he have time to, do.

In 1824 two events happened which affected the northern Troops. The first was that, as mentioned, the Marquess of Titchfield died, and the second was that the Worksop Troop, under the Earl of Surrey, became an independent Troop and renamed itself the Worksop Yeomanry Cavalry. This may have been prompted partly by the death of the Marquess of Titchfield, which may have caused the Earl of Surrey to remove his Troop from the umbrella of the Regiment so as to avoid the possibility that the 4th Duke might turn to him to command it instead. At thirty-three he was about the right age.

It is more likely that the Earl of Surrey's reasons were personal as it is known that the 4th Duke had, about that time, opposed the idea of the country having a 'Popish Earl Marshal', the hereditary appointment of the Dukes of Norfolk, and that, hardly surprisingly, had not been taken well by the Earl of Surrey. The Earl is unlikely to have been mollified had he known the 4th Duke 'meant nothing personal, I merely did my duty as a Protestant legislature'. That, and the 4th Duke's opposition to Emancipation, given the Earl of Surrey's active support of it, made a convincing pretext.

Effectively all the troops were independent because there was no overall commander.

Chapter 15

Transition 1824

All the foregoing events and the way the Regiment responded to them raise some interesting questions:

Why did the Nottingham Town Troop disband in 1808 and not reform for ten years, even though Nottingham was at the focal point for most civil disturbances?

Why did so many of the 'town' troops (for example Newark, Nottingham on reforming, and Worksop) declare themselves to be independent troops?

Why were the 'southern troops' so keen to distance themselves from the 'northern troops'?

Why did the aristocratic landowners increasingly raise troops based on their estates and not the local towns as they had originally?

Why was there no difficulty in raising troops in 1794, 1804 and 1817 to face invasion and republicanism, but none were raised in 1800 and 1811 when outbreaks of civil disorder were no less severe, but consisted of the Bread Riots and Luddism?

A troop of yeomanry, whilst acknowledged to be the best means of keeping the peace, which the vast majority on both sides wanted to see, was also seen to be too closely aligned with the aristocracy and too symbolic of the disproportionate share of power they enjoyed. This meant that when the issues were those relating to dealing with republican motivated riots and the threat of invasion, where the vast majority were in common cause with the aristocracy, there was strong support for the yeomanry and a willingness to share in the task of raising, joining and maintaining troops. Where it was a matter of the price of wheat or the automation of the manufacturing processes, where the interests of the establishment were seen to be in conflict with the interests of the people at large, there was a lack of willingness in the towns to become involved or to be seen to be in support. Often in those circumstances there was no desire to form a troop or, if one was formed, to at least make sure it was seen to be independent of and disassociated from the Regiment which was under command of leading members of the aristocracy. It is well known that, nationally, the yeomanry regiments were not popular, even though they had often been of valuable service. This explains why.

It is worth commenting that the relative instability, which seemed to surround many of the Troops of Yeomanry in Nottinghamshire, may in part be due to another reason as well as that just given, a reason that is just as relevant today. Where an organization is based on widely dispersed sub-units (as opposed to an organization based on a full unit in a single location) it is difficult for each sub-unit to guarantee a flow of succession to command it. The distances between sub-units are too great to permit the cross-posting of key officers, even if that would have been acceptable. This was most likely to occur when the individual, who had originally formed the Troop and was probably funding it, ceased to be able to commit the time to it and he was effectively the only experienced officer. These observations may go some way to answering the foregoing questions and to explaining some of the events which had taken place within the Regiment in the past.

Be that as it may, it was now the challenges of the present and the future which were of greatest importance. In 1826, as a result of the improved political climate, the Government's attitude towards independent Troops hardened and their existence became threatened unless they formed into proper regiments, rather than the much looser structures under which they had operated until then. This was not a recent change in policy, but one which, as has been seen, had existed since 1813 at least, which had been ignored in typical yeomanry style until the Government finally lost patience and made it compulsory.

The next part of the Regiment to react to the Government's policy, after the 4th Duke's partially successful effort to form the Clumber Yeomanry, were the southern troops, namely those from Nottingham, Wollaton, Watnall, Holme Pierrepont and Bingham, all situated close to Nottingham who, on 5 May 1826, paraded 311 strong in the market place at Nottingham. Afterwards, at a dinner at Thurland Hall, which had been acquired by the 4th Duke, it was decided to form them into one regiment under Captain Henry Willoughby who had hitherto been the commander of the Wollaton Troop and who went on to command the new regiment for eleven years. Thus, on 7 August 1826, the South Nottinghamshire Yeomanry Cavalry was approved by the 4th Duke as Lord Lieutenant. Its name remained unaltered till 1887 after which, as mentioned, it became known as the South Nottinghamshire Hussars Yeomanry and for consistency will be referred to as the South Notts Hussars for the remainder of this account. From the very beginning it was a strong and successful amalgamation.

One of the keys to the southern regiment's success was that Charles Arthur Herbert Pierrepont, the 2nd Earl Manvers (1778–1860), exercised significant influence over them. At a personal level he opposed Catholic Emancipation, a hot political issue in the 1820s, and to that extent would have been in opposition to the Earl of Surrey but at one with the 4th Duke of Newcastle. He was a Conservative, as was the Duke of Newcastle, but supported the Reform Bill which was a Whig measure, in which regard he would have had the 4th Duke of Newcastle as an opponent, and it is known that they were not all that close, the 4th Duke once describing him in his diaries as 'from his likeness to the animal – I call Porkus'. Earl Manvers may have concluded that soldiering in the yeomanry may have been more enjoyable if that yeomanry was a different one from that in which the 4th Duke of Newcastle served.

At the time of formation of the South Notts Hussars, Charles Evelyn Pierrepont, Viscount Newark, the eldest son and heir of 2nd Earl Manvers had recently taken command of the Holme Pierrepont Troop. He was born on 2 September 1805 and so would have been twenty-two-years-old at this time, and was in the process of graduating from Oxford with a First in Classics. Given that the Manvers family's involvement with the Nottinghamshire Yeomanry had almost always been through the Holme Pierrepont Troop, it is logical that, putting aside any conflict between the 4th Duke and the 2nd Earl, when the Regiment split the family's allegiance would be with the South Notts Hussars, even though they lived at Thoresby, in the north of the county. By 1828, when still only twenty-three Charles Evelyn is, surprisingly, shown as a major and second-in-command of the new regiment. He then resigned in 1830.

This short and unconventional career seems to indicate two things. Firstly, and self evidently, Lord Newark had little aptitude for or interest in soldiering. Secondly, the fact that he had nevertheless accepted the duty, and the regiment for its part had promoted him in the way it had, indicated that the real influence behind the formation of the southern regiment had been, as suggested earlier, his father, the 2nd Earl Manvers. This once again confirms that the enormous contribution of both he and the 1st Earl Manvers to the yeomanry in Nottinghamshire was still continuing undiminished.

In 1832 Viscount Newark married Emily, second daughter of 1st Baron Hatherton. He then became MP for East Retford, the home of his mother, the daughter of Colonel Eyre, and a seat in which the 4th Duke took a keen interest. He sat from 1830 to 1835. Given he had stood as a Whig, it must have annoyed the 4th Duke more than a little. He died without children, and before his father, in 1850. His younger brother Sydney William Herbert Pierrepont, who eventually succeeded as 3rd Earl Manvers was born in 1825 when his father would have been forty seven.

By 1827 the Northern Troops were configured as follows:

Troop or Corps	Name	Date Gazetted
Newark Troop	Captain Francis Chaplin	21/11/94
	Cornet J. Handley	22/8/08
Clumber Troop	Captain J. E. Denison of Ossington	Took command when the 4th Duke resigned. Resigned 1825
	Captain Henry Simpson of Babworth	
	Lieutenant W. H. Barrow	1/4/20
	Lieutenant S. E. Bristowe	10/11/19
	Cornet P. R. Falkner	10/11/19
Mansfield Troop	Captain Lord Frederick Bentinck	10/11/19
	Lieutenant J. Heygate	1/1/10
Worksop Troop	Captain The Earl of Surrey of Worksop Manor	24/6/19
	Lieutenant The Hon. George Monckton-Arundell of Serlby	?
	Lieutenant Sir Thomas Wollaston White Bt. of Wallingwells	5/6/24
Retford Troop	Captain J. Kirk	17/12/19
	Lieutenant Thomas Taylor Worsley	17/12/19

The Retford Troop disbanded in 1826.

2nd Duke of Portland of Welbeck Abbey. (By kind permission of Mr William Parente)

3rd Duke of Newcastle-under-Lyne of Clumber Park, Lord Lieutenant, 1794–1795. (By kind permission of Nottingham University)

3rd Duke of Portland, Lord Lieutenant, 1795–1809. (By kind permission of Mr William Parente)

Captain Ichabod Wright wearing the
uniform of the Nottinghamshire
Yeomanry, 1794. He commanded the
Nottingham Troop.

Troopers 1798
Nottinghamshire Yeomanry Cavalry
Newark Troop.

Nottinghamshire Yeomanry Troopers
Uniform, 1798. Also shown are Newark
Castle and the Royal Standard.

Charles Medows – Viscount Newark and 1st Earl Manvers of Thoresby Park. (By kind permission of the Manvers Trustees)

Captain Richard Lumley-Savile – 6th Earl of Scarbrough, of Rufford Abbey, who raised and commanded the Rufford Troop. (By kind permission of the Earl of Scarbrough)

5th Baron Middleton of Wollaton Park with his family. His heir, the future 6th Baron Middleton is on the right and second from the left is the 5th Baron's daughter, Henrietta, who married Captain Richard Lumley-Savile and whose needlework created the Royal Standard. (By kind permission of Lord Middleton)

Major the 4th Duke of Newcastle-under-Lyne who commanded the Clumber Yeomanry and was Lord Lieutenant, 1809–1839.

5th Baron Middleton as a younger man. (By kind permission of Lord Middleton)

Nottingham in the eighteenth century. Nottingham Castle and St Mary's Church can both be clearly seen. (By kind permission of Lord Middleton)

Lt Col Thomas Wildman of Newstead Abbey raised and commanded the Sherwood Rangers Yeomanry 1828–1836.

Lt Col Sir Thomas Woollaston White Bt of Wallingwells Carlton-in-Lindrick who commanded the Sherwood Rangers Yeomanry 1836–1852. (By kind permission of Sir Nicholas Woollaston White)

Nottingham Castle in Flames by the hands of the Reform Bill Rioters, 10 October 1831.

Lady Woollaston White whose needlework created the Mansfield Troop Standard presented in 1840. (By kind permission of Sir Nicholas Woollaston White)

Vere Viscountess Galway 'The Mother of the Regiment' whose needlework created the Retford Troop Standard presented in 1894.

South Nottinghamshire Hussars Yeomanry   Sherwood Rangers Yeomanry

17th Yeomanry Brigade.

Trooper's uniform of the Sherwood Rangers Yeomanry, 1840. In the background is Nottingham Castle rebuilt following the fire. Also shown is the Provincial Standard presented in 1795.

Officer's full dress uniform of the Sherwood Rangers Yeomanry, 1899. In the background is the Greendale Oak and the Retford Troop Standard presented in 1894.

3rd Earl Manvers in the full dress Uniform of the South Nottinghamshire Hussars Yeomanry, which he commanded. (By kind permission of the Manvers Trustees)

The Royal Standard, 1795 (Newark Troop).

The Provincial Standard, 1795 (Clumber Troop, Worksop Troop).

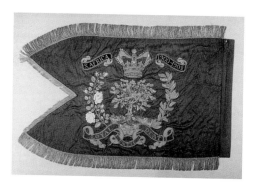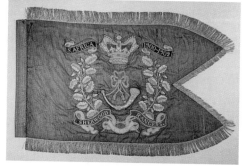

The Mansfield Troop Standard, 1840.

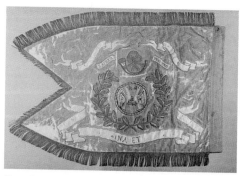

The Retford Troop Standard, 1894.

Chapter 16

The Formation of the Sherwood Rangers Yeomanry 1828

By 1828 the four remaining troops in the north of the county were still operating with no overall command, other than the over-arching input from the 4th Duke as Lord Lieutenant. There was no sign that either the 4th Duke or the four Troop Leaders were inclined to obey the outstanding instruction to amalgamate, clearly hoping to maintain the status quo. In late-1827 there had been Reform Bill riots in Arnold which had not involved the yeomanry but, in January 1828, the 4th Duke took the opportunity to use the unrest as a pretext to appeal for special permission to maintain four separate troops. This was summarily refused by the War Office with a strong injunction to form them into a new Corps.

The War Office was even more insistent this time because it had developed a fresh strategy with regard to the yeomanries which was designed, as ever, to save money. It identified those counties perceived to be the most unstable and to disband all yeomanries, other than in those counties. This resulted in the official survival of only twenty-two regiments or corps compared with the fifty-one regiments or corps that existed in 1798, when the Napoleonic threat was at its highest.

In the case of Nottinghamshire, the Government's insistence that the northern Troops be amalgamated to form a new corps meant there would now be two corps in the county. Thus Nottinghamshire became one of only four counties throughout the United Kingdom where more than one regiment or corps was retained; the others were Glamorgan (the only one with three), Shropshire and the West Riding of Yorkshire. The Government must have felt that Nottingham had been one of the most unsettled cities in the country; alone it had, in the past, needed effectively a whole regiment to keep the peace and it was presumably assessed to still need that. Government then, seemingly, decided that there was also a need for cover in the northern part of the county, particularly in respect of Mansfield, which had previously also been a hot spot, and this required a second regiment.

Before the 4th Duke could carry out the instructions he had been given, he needed to find someone to command the Corps. The obvious place to look was amongst the existing Troop Leaders. He found that the Earl of Surrey had, in the new year, taken the decision to disband the Worksop Troop. The reason is not known, but he was, of course, by then a Privy Councillor and clearly had political ambitions which would be realized in the near future. The remarkable Captain Francis Chaplin who had commanded the Newark Troop virtually since it was raised in 1794, and without ever allowing it to disband or merge, was due to retire, and his successor, Captain John Handley, would obviously be too inexperienced. The Hon. George Monckton-Arundell had only recently taken command of the Clumber Troop and so also lacked the necessary experience or seniority. Finally, Captain Lord Frederick Bentinck, who would have been ideal, had a terminal illness from which he would die later that year and had retired, handing over the Mansfield Troop to Lord Byron's friend, Thomas Wildman. There clearly lay the answer.

In 1816, following Waterloo, Wildman had purchased a majority in the 2nd West India Regiment. Later he had transferred to the 9th Light Dragoons but had now left the Army. He was also at some stage the aide-de-camp to the Prince Augustus Fredrick, Duke of Sussex, the sixth son of George III. Apart from being Captain General and Colonel of the Honourable Artillery Company, the Duke was the only son of the King not to pursue a military career, his interests lying in the arts. Thomas Wildman had considerable personal wealth based on assets in the West Indies acquired in interesting and, it is said by some, controversial circumstances.

In outline Thomas Wildman had a younger brother, James Beckford Wildman (1789–1867) who inherited Chilham Castle in Kent from their father James Wildman. Whilst Thomas Wildman was serving in the Army James Wildman created business interests in the West Indies which included lending money to plantation owners. Given the fair and generous way he was treated by his brother, Thomas Wildman may well have been a sleeping partner. One such owner was William Thomas Beckford (1760–1844), said to be the largest owner of plantations in the Caribbean, which were of course at that time, run by slave labour. He was also said to be the richest commoner in England with an estimated annual income of £100,000 (£10 million today). He was an art collector on a grand scale and, more controversially, became known to be bisexual in a scandal which ruined him socially. (By contrast, to be a slave owner at that time was acceptable.) It is said that Beckford failed to repay loans in favour of James Wildman secured on Beckford's Quebec Estate, which at 800 slaves (200 was the average) was said to be the largest and most profitable sugar plantation in Jamaica and that, as a result of the non-payment, the estate became owned by James Wildman.

Thomas Wildman's wealth was said to be derived from the Quebec Estate and in December 1817 he purchased Newstead Abbey for £94,000 from his old school friend, Lord Byron, who was having financial difficulties and had been trying to sell the Abbey since 1812.

Newstead Abbey had been the Byron family seat for almost four centuries, so it was a sad occasion for the family. Of the sale, Byron's half-sister Augusta said Wildman had 'soul enough to value the dear Abbey ...'. She was right. He spent £100,000 restoring it, hiring the architect John Shaw to make improvements. He also amassed a large collection of Byron memorabilia there. As a result he is credited as the man who saved Byron's home. Thomas Wildman also built what he called the Duke's Tower in the grounds of Newstead in honour of a visit there by the Duke of Sussex. By this time Lord Byron's stardom had both risen and fallen, establishing him as a man who was either adored or loathed, and so being so closely associated with him was not necessarily a wise thing, which indicates that their friendship must have been a very strong one. Lord Byron died in 1824.

Thomas Wildman, now aged forty-one, may not have had connections with any of the great county families, but he was clearly a man of standing in the county with all the experience needed to command the Corps. Indeed, the fact that he had no such connections may have been seen as an advantage since he came with no local prejudices and an open mind. The 4th Duke offered him the job and he agreed to take it on. He took command in June 1828 and was designated Major Commandant. He set about creating what was, in effect, a new corps of yeomanry with tact, firmness, diligence and skill. This comes across very clearly from the pages of his letter book covering the first two years of his command, which has survived and is in the regimental archives of the Sherwood Rangers Yeomanry.

One assumes that the subject of a title for the new Corps must have come up when the 4th Duke and Wildman discussed his appointment. Major Wildman suggested Sherwood Forest Rangers, but this was almost immediately changed to Sherwood Rangers, and adopted forthwith. Indeed Major Wildman's letter book, which records the very first correspondence relating to his command, is entitled 'Sherwood Rangers Yeomanry'. It is likely that, out of courtesy at least, Sir Thomas White's permission was sought. If so it would have been his only part in the formation. The reason for the choice of name was that they needed to find a distinctive identity for the Regiment in its new form. There is some evidence that the old title of the Nottinghamshire Yeomanry was being used informally to describe the four, now three, independent troops, but there may have been objections to that because the title was too similar to that of the southern regiment.

It might have been expected that, if the Regiment was to be named the Sherwood Rangers, the uniform would have been rifle green, but there is confusion because, in Wildman's letter book, dark blue or indigo was selected, to be based on the uniform of the Derbyshire Yeomanry. However a contemporaneous letter from Wildman in the 4th Duke's papers talks of green. There is no other evidence which clarifies the matter and there is certainly no evidence that either the 4th Duke, or Wildman, was thinking down the line of incorporating any of the identity of the White family's unit other than the name, which had been dormant for twenty years.

It was decided that the new Corps of Yeomanry would, as expected, consist of an amalgamation of the Newark, Clumber and Mansfield Troops, together with the addition of the Newark Troop's Brass Band. None of them could be described as strong. The best of them was the Newark Troop, but even they were not sufficiently strong in privates to avoid the necessity of showing the band as part of the basic troop strength rather than as surplus to establishment. Neither the description of Trooper or Yeoman, seemingly, had been adopted at that point to describe those in a cavalry regiment who did not hold rank.

The Mansfield Troop did not, of course, have a Troop Leader, and had a strength of forty Privates against an establishment of seventy. The weakest of all was the Clumber Troop which had a strength of fifty-two Privates but, more worryingly, had recently been described by Captain John Evelyn Denison MP, its former Troop Commander, as being in a state of 'utter inefficiency and dilapidation, having not paraded since 1826'. Their state of dilapidation is hardly surprising given their uniforms had not been renewed for ten years. The reason was poor administration which had resulted in the uniform allowance available annually from the War Office on a per capita basis not having been collected by the troop for a number of years.

The establishment of the new unit, presumably specified by the War Office, was divided between the three troops as follows:

Major Commandant	1
Captains	3
Lieutenants	6
Cornets	3
Adjutant	1
Surgeon	1
Sergeant Majors	3
Sergeants	10
Corporals	10
Privates	210
TOTAL	248

The Troop Leaders and the strength of their Troops against a Troop establishment of eighty-one at the time of handover were:

Newark:	Captain John Handley	72 (including the Band)
Clumber:	The Hon. George Monckton-Arundell	53
Mansfield:	Acting: Lieutenant L. Rolfe. Captain Trebeck appointed later	70
TOTAL	Including 3 RHQ appointments	198

Major Wildman took the new Regiment to Camp for the first time for 'six days exclusive of days marching' on Monday 13 October 1828 in Mansfield, this being the town, other than Nottingham, from which trouble might be most expected. The Regiment assembled on the racecourse before moving into billets in the town itself, an event described in the Nottingham Journal in the following terms:

> ... and about 4 o'clock, preceded by an excellent band, the whole marched into Mansfield in grand military style. Some hours before their arrival, the streets were crowded with people anxious to witness the entry of the troops. The whole composed a fine body of men, well mounted, and gaily attired. The town is much thronged, and the theatre reaps some advantage. On Monday evening the play of John Bull (bespoke by Mrs Wildman) attracted a splendid house; and on Wednesday The Rivals (bespoke by the Colonel and the officers of the troop) again occasioned an overflow.

Between taking command and Camp, a period of a mere fourteen weeks, Major Wildman succeeded in recruiting the Regiment up to full strength so that it attended Camp less only one lieutenant, one adjutant, three cornets and three privates, although the Newark Troop was still counting the band against their establishment at that stage. He also, remarkably, designed, had made and issued uniforms for all ranks in time for Camp. This included all accoutrements including pouch belts, sabretaches and saddle cloths, but not saddles and bridles, or, indeed, horses which the yeomanry continued to supply themselves. Camp cost £740 5s 8d and followed a pattern of basic mounted and dismounted drills which increased in their range and difficulty towards a final parade and day of manoeuvres on Mansfield racecourse, which were 'commendably executed'. The Regiment, in its new form, was on its way, and the achievement in so short a time reflected well on all, but none more so than on Major Wildman himself. His achievement is all the more noteworthy given that on 5 September 1828, in the middle of his preparations for camp, he had to deal with what might be described as a diversion when he received a letter from the Ordinance Office asking him to account for all weapons ever issued to the troops making up the new unit since 1803, a quarter of a century before. The weapons were of the following categories: all Carbines with bayonets and scabbards and rammers; Pistols with rammers; Sabres with scabbards; Belts; Knots; Trumpet with string; Bugles with string.

This will strike a chord with all who have had military weapons on their signature, but few would have been able to remind the Ordinance Office that, on this occasion, the individuals with these items on their signature did not include him, but did include a Duke, who was also the Lord Lieutenant and a Knight of the Garter and the heir to a Dukedom who was also a Hereditary Earl Marshal of England (elect). The weapons seemed to emerge from various armouries in the bowels of various stately homes and, remarkably, he was, in due course, able to submit a return for the correct numbers of each item, which leads one to suspect that those were not all the weapons secreted in such places.

One of his best innovations was to recruit a permanent sergeant for each troop starting with the Clumber Troop. He did so by finding a suitable individual, paying him an honorarium and persuading some landowner associated with the troop to give the man a part-time job as well. All who understand part-time soldiering know how important it is to have some form of permanent infrastructure. Running a volunteer regiment was once famously described by a frustrated training major during the Cold War as 'like trying to command a regiment on permanent block leave'.

No sooner was Major Wildman's first period of Permanent Duty, or Camp, complete than he began preparing for the one for the following year. The first priority, with more time for thorough planning than he had enjoyed the previous year, was to establish the best time of the year for a yeomanry regiment to gather for training. He made his enquiries and decided 'I find upon enquiry that it seems to be the general wish of the Agricultural Members of the Corps to go out on Permanent Duty in early May', thus setting a pattern destined to be followed for well over a century until the Cold War autumnal exercise season caused it to change.

In 1829 he obtained the permission of the Lord Lieutenant for the Regiment to attend its Permanent Duty at Newark from 4 May for six days, excluding marching days. For that camp he introduced another valuable innovation, which was to arrange for each troop to parade for his inspection during the month before attending the period of permanent duty so that he could identify shortcomings and, where possible, correct them before the unit assembled, rather than when they had already done so.

In 1830 he sought the Lord Lieutenant's permission for the period of Permanent Duty to be in May again, this time near Retford, in a suitable location on the Clumber estate, setting up a recurring cycle between Retford, Newark, Mansfield and Worksop, the towns from which unrest needing the intervention of the yeomanry would most likely originate, which will be referred to later; this pattern would endure for the rest of the century. Every year he introduced a new improvement in the lead up to the annual period of Permanent Duty. This time it was new bridles made and issued to the entire Corps.

From this point reference to "the Regiment" is to the Sherwood Rangers Yeomanry.

Chapter 17

The Reform Bill Riots 1830–2

As mentioned, civil unrest had subsided during the 1820s as that decade produced increasing prosperity which took the edge off the earlier discontent. Time had also brought changes in the country's political leadership. By 1828 ill health and premature death had brought an end to the political careers of Liverpool, Castlereagh and Canning. This had also triggered a brief period of political unrest which culminated in the appointment of Wellington, the only survivor of the previous administration, to be Prime Minister, and Robert Peel, a fast emerging political giant, to be Home Secretary. They both represented the status quo.

They first had to address the question of Catholic Emancipation which had been at the top of the political agenda since the end of the war and which, as has been mentioned, had also strained relationships in the Dukeries. In 1829 the Government agreed to accept the policy in order to remove the risk of civil war in Ireland, and persuaded the King to support it to prevent the Government from falling and being replaced by the less-appealing alternative of one led by the Whigs. As mentioned, the Earl of Surrey was quick to take advantage of the change, whilst the 4th Duke did not omit to make his views known.

George IV died in June 1830 and was succeeded by William IV.

Now Reform was swept back onto centre stage by yet another revolution in France overthrowing the newly-appointed Bourbon monarchy. This, and other issues to which it was related, triggered such unrest in the country that, in August 1830, a general election was precipitated but was indecisive. Reform split politicians across party boundaries, but Wellington and Peel were implacably opposed, and when radical Tories, as a result, voted with the Whigs, now decisively in favour of Reform, the Government was defeated. William IV called for Earl Grey, the Whig leader, to form a Government which he did, bringing the Whigs back to power for the first time, apart from one short interlude, in forty-seven years.

Such was the unrest in the country and the Commons that whether or not the new Whig Government wished to tackle Reform they had no choice; any sign of prevarication and they would have faced defeat before they had truly taken office. Nottinghamshire in general and the 4th Duke of Newcastle in particular were not immune from the unrest, and precautionary meetings of the Lieutenancy were held, attended by the 4th Duke as Lord Lieutenant to make contingency plans. Whilst none of the unrest was sufficient to involve the Regiment at that stage the Times reported on 9 December 1830 that in Newark

> the Sherwood Rangers have received orders to be at one hour's notice; farmers in the neighbourhood have been sworn in as special constables ... A hovel at Bennington was fired on Saturday night; various other excesses have been committed in the neighbouring towns in Lincolnshire etc.

In March 1831 the Whig Government moved their Reform Bill for the first time. The Government could not risk producing a half measure, and it contained a list of over a hundred 'rotten' or 'pocket' boroughs for abolition, a number being those of the 4th Duke, to be replaced with new constituencies for London and the northern and midland industrial cities. As can be imagined, this was hugely popular in the country, which stiffened the Whig resolve to prevail against the outraged Tory opposition, who argued that the seats under threat of abolition were private property and, as such, the Bill was a recipe for revolution. The Whig majority, relying, as needs must, on the support of other minorities, was far from secure, and after a desperate fight the measure was defeated in the spring of 1831. This caused such a fear of unrest locally that the Sherwood Rangers were called out for a period in June 1831 which, locally, had the effect of calming the situation.

Amid a storm of protest and outrage in the country, Earl Grey asked the King for a dissolution which was wisely granted in April 1831 and resulted in a general election on the single issue of Reform. The Tories were defeated heavily, the Whigs and their allies winning a landslide victory of 136 seats. With a majority in the House of Commons now secure, the battle switched to the House of Lords where it was fought with all the same passion against a highly charged atmosphere in the country. The final vote took place on the night of Friday 7 October 1831 and once again the Tories prevailed, the 4th Duke prominent in the process. The question now was that of the Peers versus the People. Angry crowds gathered outside the Houses of Parliament. The 4th Duke's own outright resistance to Reform, and his well-known use of the present system marked him out particularly, drawing crowds outside his London house in Portman Square, which broke all the windows, having first smashed the shutters, and only narrowly failed to break down the front door.

There was worse to come. News of the vote reached Nottingham on Saturday 8 October and immediately heavy rioting broke out and continued on and off throughout the weekend. Initially the unrest was contained by judicious use of the South Notts Hussars and a small detachment of the 15th Hussars. On Monday a poor decision was taken by the Town Clerk to dismiss the Wollaton Troop which had, until that point, been stood to at his disposal. This enabled the rioters to break away and not only to sack Colwick Hall, but to set fire to Nottingham Castle, no less, which towered above the town, and which belonged to the 4th Duke, totally destroying it. The building destroyed was not the medieval fortress which had been demolished some years before, but a replacement which had the look of an Italian palazzo, so the mob may have done one and all a favour. Due to its commanding position over the town, initially its charred remains and latterly the indifferent and equally inappropriate building which eventually replaced it, has left a prominent monument to one of the great

Nottingham Castle.

political struggles of this nation's history. The Wollaton Troop had not stood down on being dismissed by the Town Clerk but had withdrawn to guard Wollaton Park which was also threatened by the mob. However, assisted by two of the estate's gamekeepers locally known for their marksmanship, they were successful in protecting that beautiful house.

On 12 October two Troops of the Sherwood Rangers, almost certainly the Newark and Mansfield Troops, were called out for duty in Mansfield and had to deal with disturbances at Plumtree and in Mansfield itself on 13 October. The 4th Duke also heard that there was a threat to his home, Clumber, and returned immediately from London taking care not to disclose his identity as he travelled. On arriving home on 13 October he found that his eldest son, Lord Lincoln, a future Commanding Officer of the Regiment, but then not serving in it, had laid out comprehensive defences. To do this, he had been supported by the remaining Troop of the Regiment, presumably the Clumber Troop and people from the estate. All were stood down after two days as the heat went out of the situation.

In October 1831 Thomas Wildman received a letter from Lord Melbourne the Secretary of State or Home Secretary which stated

The Major General Commanding the Northern District having reported to me the very excellent conduct observed by the Sherwood Rangers called out under your command in aid of the Civil Power during the recent disturbances in the neighbourhood of Mansfield. I have great pleasure in expressing my best thanks to you and to the officers, (non commissioned) officers and privates composing these troops for the steadiness and zeal displayed in the difficult duties they had to perform and I shall not fail to report to His Majesty the useful services rendered by these troops to the Civil Authorities in repressing the disturbances and in preserving the public peace.

The Sherwood Rangers were also thanked by the Magistrates of the County.

This indicates how well the Sherwood Rangers had done. In addition during the same year Colonel Wildman was promoted to lieutenant colonel, presumably in further recognition of the Regiment's performance during the riots.

In December 1831 the Reform Bill was introduced once more. Again it passed the Commons, this time with a majority of two to one, and again it went on to the Lords only to be defeated by forty-four votes. The Government watered it down further but the House of Lords was still opposed, and Lord Grey appreciated the likelihood of another rejection. He asked King William to create another fifty Whig peers to ensure its passage, but the King, in a move which carried strong disapproval nationally, refused, at which point Lord Grey resigned from his post as Prime Minister and the Government fell.

The High Sheriff of Nottinghamshire at the time was Henry Machin, now of Gateford Hall, who had served in the Clumber Troop. These events coincided with a number of Assize Dinners being hosted by the High Sheriff, which were perfectly routine affairs, but emotions were running very high and, in that atmosphere, an allegation appeared in the Times that at one of the dinners Henry Machin had toasted only the Queen and not the King. If true, this would have represented an insult, presumably reflecting his disapproval of the King's decision. This triggered a critical letter in the local press from a Mr H. S. Wake, an attorney of Worksop and known to Henry Machin. Mr Wake was likely to have been the son of William Wake, the one-time agent of the Duke of Norfolk on the Worksop Manor Estate. The Wake family had already founded the firm of solicitors in Sheffield now known as Wake Smith on the back of their relationship with the Duke. But Mr H. S. Wake was not part of that firm. The allegation was not true and Mr Wake's intervention triggered an angry response from the High Sheriff and Mr Wake backed down. This story illustrates how heated matters had become.

Meanwhile the King's decision triggered widespread and significant riots throughout the country, particularly when he followed it up by asking the Duke of Wellington to form a new Tory Ministry which could only have resulted in the Reform Bill's defeat.

In Nottinghamshire many people were so outraged that the High Sheriff received a petition asking him to hold a public meeting to give the people of the county an opportunity to make their views known to the King. He was no doubt chosen because at that time the High Sheriff was the King's representative in the county and the senior office holder as well as the only one who outranked the Lord Lieutenant who would have been an alternative to receive such a petition, but not in the circumstances which prevailed at that point in Nottinghamshire. The office of High Sheriff had in the past been a controversial one, since the sort of power that it carried created many temptations, but by the nineteenth century it had become strictly apolitical so that Henry Machin would not have welcomed the risks to his position that such a petition represented.

Nevertheless, as another clear indication of the strength of feeling generated by Reform, the High Sheriff agreed to act on the petition and convene a meeting which was duly held at the Moot Hall in Mansfield on 25 May. In the event, within the space of the few days between it being called and held, the Duke of Wellington had ascertained that there was no support for him to form a Government and stood aside, making way for the King to invite Lord Grey once more to form a Government and create the necessary additional peerages; thus victory for those in favour of Reform was now assured.

In a county the size of Nottinghamshire the expectation, if fifty peerages were to be created nationwide, would be that up to two loyal Whigs would be ennobled and so there was bound to be speculation as to who might be chosen. Between the meeting being called and held, the Tories in both the Commons and Lords capitulated, thus ensuring that the Bill would be passed without the need for the creation of any additional peers. The meeting therefore turned into a celebratory event and overflowed outside where people were marching and bands playing. Those who attended were, of course, strongly in favour of the passage of the Bill. Amongst those present were G. H. Vernon MP, Colonel Wildman and three members of the Foljambe family who were all Whigs, and many others. All the speeches were for Reform and among those who spoke was Colonel Wildman. One might question his judgement in doing so at all, because one would expect the Commanding Officer of one of the regiments whose role it was to keep law and order in the county to be apolitical whilst in that position.

In his speech he made some stark criticisms of the system that was being changed:

Friends and men of Nottinghamshire I come forward most cordially to join in the universal acclamation of triumph for the victory we have achieved ... I was one of the first to put my name to (the) requisition and most warmly and sincerely now do I congratulate you and the people of England ... for the change which has saved us from being handed over to a faction which has shown an enmity to all Reform ... What has the House of Lords ever done for the House of Commons? They have never given up anything but talk of Reform as an unconstitutional measure. But did it never occur to their right honourable intellects that the interference of a peer in returning a member to the House of Commons is prohibited by the Law? Did it never present itself to their imagination that for a Peer to send his nominee to that House and thereby secure a control over the supplies and taxes to be levied on the people was at direct variance with every principle of the constitution? No the system worked well for them for they could put their hands into the pockets of the people without any delicacy of sentiment ... For myself I am proud to belong to the people – I always considered myself one of them; and so help me God if the sky was to rain Coronets I would not stoop to pick one up! (Cheers).

One might be excused for thinking that this was a little rich coming from one who owed his fortune to slavery. More to the point he did not mention by name the 4th Duke of Newcastle, effectively his immediate superior from a military point of view, he might just have well have done. In addition, instead of merely confining himself to constructive arguments concerning the merits of Reform as a route to democracy, he attacked with venom, and most eloquently, the ethics of those who opposed Reform, which could hardly be a more provocative stance. It is hard to see that there would not be consequences for his relationship with the 4th Duke and, therefore, for the Regiment as a result, and so there were to be.

The meeting ended triumphantly but peacefully and on 7 June 1832 'the Great Reform Act' became law. No doubt the King was relieved that the affair was finally over but one suspects no more relieved than his loyal subject the High Sheriff of Nottinghamshire.

The passing of the Reform Act 1832 significantly undermined the 4th Duke's electoral influence. He lost his four nomination seats in Yorkshire, and patronage and interest in six boroughs. Thereafter his activities were confined to Nottinghamshire where he was still able to return a member each for Newark and East Retford, and at least one of the four county members. The 4th Duke's opinions were not altered, and given the cynicism and apathy with which our fully democratic system is viewed these days and the lack of integrity with which modern politics is perceived to be conducted, who can deny he at least had a point, even though in the wrong. He continued to exert his electioneering influence wherever possible until the time of his death.

One further event of significance to both Nottinghamshire and the Regiment took place in 1832. No doubt to the relief of Nottinghamshire, 'Black Jack' Savile succeeded the 6th Earl of Scarbrough, who had raised the Rufford Troop, to become the 7th Earl and in the process inherited Sandbeck Park and Lumley Castle. Although he had a younger brother he did not, under the terms of his uncle's will, forfeit the Rufford Estates on succeeding to the Earldom, although he did move out of the county to take up residence just across the border at Sandbeck. Maybe the youngest brother had not been born when his uncle's will was made. As a result, the Earldom of Scarbrough, as owner of Rufford, now had firm links with Nottinghamshire. The Earldom dates from 1690, but the

Lumleys can trace their origins back much further, to Northumberland before the Conquest. Like the Howards, their fortunes have been heavily dependent on, and closely aligned to, their relationship with the Monarch of the day. Their service has been of great loyalty, mainly as soldiers, but in other capacities as well, continuing right up to the present day.

They have performed this service with distinction, being, on many occasions, in on the making of both kings and history. It would not be entirely right, however, to use the words 'great loyalty' in this context without qualifying them. Although loyalty was invariably the family's aim, it was sometimes not easy to decide to whom to be loyal. Like the Howards, it was inevitable that the Lumleys occasionally found themselves 'riding the wrong horse'. As a result, one member of the family served a spell in the Tower of London whilst another was beheaded for being part of a treacherous rebellion. However, like the Howards they always seemed to recover their lands and their status following these setbacks.

Chapter 18

The Chartists 1836–50

In 1836, Colonel Wildman resigned from his command of the Regiment in protest. There are two explanations concerning the cause of the protest, both involving the 4th Duke as the other party. The popular choice for the cause of the row is contained in a twentieth-century newspaper article based on the recollections of Quartermaster Sergeant Major Ironmonger (of whom more later) which states

> the political excitement over the Reform Bill of 1832 led to differences of opinion between the Duke of Newcastle and Colonel Wildman who was in command of the Rangers. The result was that the Colonel resigned in high displeasure, and the regiment narrowly escaped being disbanded for he refused to hand over the funds amounting to £1100 to his successor, and sent the money to the War Office.

It is clear that 4th Duke's strength of feeling on the subject of Reform was matched by Colonel Wildman's, and an irreparable breach in their relationship may have been opened by the public utterances of Colonel Wildman, but it is unlikely that this was the only pretext for Colonel Wildman's resignation because there is a gap of four years between the two events. The more likely trigger, fuelled no doubt by their poor relationship, is the second, namely that the 4th Duke refused to appoint a number of officers whom Colonel Wildman had recommended. Wildman's command had been characterized by a direct and pragmatic approach, and this may well have concerned the 4th Duke when it came to officer selection. What may well be true in relation to the first report was that Colonel Wildman did send the money to the War Office and, undoubtedly, as will be explained, there was talk of amalgamation with the South Notts Hussars shortly after this had happened.

Whatever the reasons for Colonel Wildman's resignation, they do not detract from the fact that he had laid strong foundations and left the Regiment better in every way than when he took command. However, failing to maintain cordial relations with the 4th Duke and his conduct in sending the Regimental funds to the War Office were not in the best interests of the Regiment. He continued to live at Newstead Abbey until he died in 1859. During that time he served for a time as the Provincial Grand Master of the Nottinghamshire Masonic Lodge and was a magistrate. Unfortunately for Thomas Wildman his West Indian estates from which he had derived his considerable wealth suffered an enormous fall in value towards the end of his life. Happily, the cause of that fall was the ending of slavery. As a result, on his death, Newstead had to be sold.

Colonel Wildman's remains are buried in Mansfield Cemetery where he is commemorated by an ornate monument, having stone lions at each corner, which was erected by his widow. The inscription on it includes the following sentiments:

Colonel Thomas Wildman born with wealth, but preferred the service of his country to a life of ease. At an early age he entered the army ... He was a gallant soldier, upright magistrate, a liberal landlord and a poor man's friend.

He was succeeded in command of the Sherwood Rangers by Sir Thomas White, the second Baronet, now aged thirty-five, who was, no doubt in some small part at least, inspired to do so by the Regiment's name, but also because the 4th Duke was keen to recruit him. He had a high opinion of him, writing of him in his diaries 'He is an excellent man'.

Following the untimely loss of his first wife in 1825, Sir Thomas, in 1827, had married Mary Euphemia Ramsay, the eldest daughter of William Ramsay of Golgar, a distinguished Edinburgh banker, who later lost his fortune when his bank failed. She was Sir Thomas's late wife's second cousin and a year older than him. It is said of her that 'she was as universally admired for her great beauty and her charm of manner as she was beloved for the even sweetness of her temper, which endeared her to all who knew her'. He had two sons and three daughters from his second marriage, the eldest of whom was his heir, also Thomas Woollaston White. To convey an idea of the nature of the man there is a story told that, on an occasion during an election in Newark, probably during one of the Reform Bill elections, there was a riot and the magistrates consulted Sir Thomas about what should be done. He advised that troops who were billeted some distance from the borough should be called out, but nobody could be found to take the order to the unit. Sir Thomas offered to do so himself. However, when he sought to cross the Trent, he found the bridge guarded by a dozen of the rioters, who threatened to throw him over the parapet. Sir Thomas was a powerful man of six feet in height. He sized up the situation and replied, 'I have no doubt you can throw me over, but I shall take care that at least one man goes over with me'. His determined manner called their bluff and made them afraid to touch him, and they allowed him to go on his way without being stopped.

Sir Thomas had served as High Sheriff of Nottinghamshire in 1833, succeeding Henry Machin. At that time the office was resplendent with ancient tradition, and he and his entourage are described, when sallying forth to meet and escort the High Court Judges selected to sit in Nottingham, as follows:

> He rode out with his javelin men, all of whom were his own people, attired in the green and silver livery of Wallingwells, and his trumpeters with silk banners painted with his coat of arms of twenty-eight quarterings, he himself mounted on a skewbald horse caparisoned in the crimson velvet and gold trappings of the great Duke of Marlborough, with his coach and four following to bring in the judges.

The trappings referred to were a family heirloom and had been given to an ancestor who had been a close colleague of the Duke.

At the time he assumed command as Major Commandant, he was already in command, as a lieutenant colonel, of the Royal Sherwood Foresters or Nottinghamshire Regiment of Militia, having taken up that appointment, in the same year that he served as High Sheriff. Notwithstanding his appointment to command the Yeomanry, it is said he continued to command the Sherwood Foresters for many years, although there must be some doubt about that. The Regiment had acquired a commanding officer who was a stylish and charismatic personality as well as being a good leader. He was described when in command of the Regiment as 'scrupulous and methodical'. He also published some carefully prepared standing orders, which were 'exemplary in their thoroughness'. He was, therefore, clearly, as keen and able an officer as ever his father had been. Following his assumption of command the officers were as follows:

Rank	Name	Date Gazetted
Major Commandant	Sir Thomas Woollaston White	26/2/36
Captains	Samuel William Need (later Welfitt) future CO	3/11/35
	Francis Hall	26/2/36
	Thomas Spragging Godfrey	2/5/36
Lieutenants	Robert Caparn	20/5/30
	Godfrey Tallents	2/5/36
	George Hutton	2/5/36
	Lord William Pelham-Clinton	2/5/36

Rank	Name	Date Gazetted
Cornets	Charles Colville	14/3/36
	Robert Hall	12/5/36
	John Ingall Werge	3/12/36
Surgeon	James P. Lacey	25/9/29

This list of officers bears several signs that the row between the 4th Duke and Colonel Wildman over officers had been a big one. Firstly, almost all, not just Sir Thomas and even the captains, had been gazetted during 1836, and so had been brought in to replace the previous incumbents, of whom no records remain. Secondly, one of the new officers was Lord William, the Duke's fourth son, born in 1815 (He went on to a career in the diplomatic corps). Thirdly, the Duke's agent in Newark was called Tallents, and the likelihood is that Godfrey Tallents was a member of his family and, incidentally, the first mention of a family destined to have a great deal to do with the Regiment during the next century or so, producing a commanding officer in the process.

We know that Thomas Spragging Godfrey was the manager of the bank in Newark where, it is likely, the 4th Duke banked and there is the following anecdote about him. Whilst he was the commander of the Newark Troop he was a zealot when it came to recruiting and was not averse to using his influence as bank manager to persuade his customers to sample the joys of yeomanry soldiering. This was not always well received by the customer but on the other hand it was hard to resist if you needed a loan. One such customer was the village butcher in Balderton:

'When shall you want me?' said the butcher in response to the Captain's approach "I'm just going to 'ave t' cart done up I wouldn't like t' come out shabby' [No doubt the cart in question was his offal hides and skins cart which would have been a gory and malodorous form of conveyance.]

'Cart! What do you want a cart for?'

'I'm a really good soldier, Sir, in a cart, I shall be forced t' 'ave t' cart.'

End of conversation.

Not only was 1836 the year when Sir Thomas assumed command but, sadly, it was also the year in which Lieutenant Colonel Anthony Hardolph Eyre died. Having retired from active involvement in politics in 1812 he had devoted himself to county affairs, serving as a Justice of the Peace for many years. Following his death his estates were divided between his younger daughters.

Firstly, Frances Julia Harcourt Vernon inherited a life interest in the estates at Grove and Headon jointly with her husband Granville Harcourt Vernon with remainder to their eldest son Granville Edward Harcourt Vernon. This bequest represented the majority of Colonel Eyre's estates. Secondly, Henrietta Eyre, in whose case the brother of her late husband, John Hardolph Eyre, inherited Rampton. She was married to Henry Gally Knight MP, her second husband, and, in common with her eldest sister, Lady Manvers, was well provided for and so the fact that the main beneficiary was the middle sister's family was not controversial.

Under Sir Thomas's command, periods of permanent training began to acquire the stylish and imaginative embellishments which became a part of such occasions for the rest of the century. This reflected a continuation of the yeomanry tradition of not just hard work, but hard play as well. For example, he commissioned bronze medals, inscribed on the front with a mounted cavalryman surmounted by the motto 'Loyal until Death' and on the reverse with the words 'For long service and good conduct in the Sherwood Rangers Yeomanry Cavalry'. Another example is that Yeomanry Cavalry Races were held at Worksop on 12 May 1838. Presumably this marked the climax of the annual training as horses 'had to be ridden eight days in the Corps in May, 1838 ...' to qualify to be entered. There were two flat races of one-and-a-half miles and a hurdle race. Again, there is an advertisement in existence of a Yeomanry Theatrical Show at Newark just before Christmas 1838.

As was observed by the 4th Duke, Sir Thomas was also a thorough commander and there is evidence that he introduced new carbines and pistols for use by the Regiment. When he took command the carbines and pistols issued to the Regiment were flintlocks from the Napoleonic era and not the modern pin-firing percussion weapons. They were so obsolete that even the technology of sporting shotguns, which members of the Regiment would have been using to shoot game, had overtaken them since shot guns were now percussion fired. He therefore took the decision to purchase percussion carbines and pistols, made to the Regiment's own unique design based on the design of contemporary shotgun firing mechanisms as described and probably ordered from Birmingham. Its calibre was .66-inch. Starving units of the basic necessities to do their job is clearly a skill which has been fiercely nurtured and developed down the years by the War Office and its successor the Ministry of Defence. Eventually, the War Office caught up and introduced the 1844 pattern Yeomanry carbine.

Events now took place which at first looked as though they would threaten the Regiment's very existence but which, in the end, became so serious that they gave the Regiment the opportunity to establish its reputation.

William IV died in 1837 to be succeeded by Queen Victoria, aged eighteen. Her succession coincided with the start of a severe economic downturn, as severe as in the period following the end of the Napoleonic Wars, and, as a result, the Government made a second major cut in the size of the yeomanries paid from public funds. This time few landowners were prepared to take on the funding of troops privately and therefore the cuts became permanent. There was a move to amalgamate the Sherwood Rangers, still at that stage consisting of three troops – Newark, Clumber and Mansfield – with the South Notts Hussars. In part this may have been triggered by the fallout from the row between the 4th Duke and Thomas Wildman as claimed by Quartermaster Sergeant Ironmonger. The amalgamated regiment was to be called the Nottinghamshire Yeomanry Cavalry. However, true to form, it did not take place because of opposition to it by those concerned, in particular, one suspects, the 4th Duke, but also the Duke of Portland, because of his concerns for security in Mansfield. Presumably the Government was willing to indulge this resistance because of the continuing potential for unrest within the county.

The economic downturn, in combination with the unpopular Poor Laws of 1834, had further serious effects. It stirred renewed political unrest. Once again it was the Radicals from the industrial heartlands who were at the centre of it and, once again, the fundamental issue was republicanism. This time, Reform having been addressed, for the time being at least, it needed a new name and a different agenda. The name was Chartism and the agenda was set out in a 'Peoples' Charter' published in May 1838 by a group of working class leaders. Their Charter advocated annual parliaments, universal male suffrage (the Reform Act had only extended the vote to male property owners), equal electoral districts, the removal of the property qualification which applied to Members of Parliament, a secret ballot, and the payment of MPs. It is regarded by some as the birth of socialism.

As can be seen, when viewed with hindsight, the demands were mostly reasonable. They could not be achieved, however, without the support of both a major political party, and the middle classes. The Radicals not only shunned this support but, in any event, forfeited any chance of securing it, because of the extreme nature of both their rhetoric and their actions. These included, for example, the secret arming of their hotheads. The intensity of their cause ebbed and flowed with the state of the economy, as did its impact on the Regiment.

The first, and significant, involvement of the Regiment came in May 1839. It was preceded, however, by a most remarkable sequence of events at the centre of which was the 4th Duke of Newcastle, as controversial as ever. Rumours started in January 1839 that Chartists in Nottinghamshire were arming, the two centres for trouble being Nottingham and Mansfield. These rumours triggered a meeting on 20 February 1839, on the initiative of the 4th Duke of Portland and the 2nd Earl Manvers, with the 4th Duke of Newcastle. The concern of the Duke of Portland and Earl Manvers was directed at unrest nationally and how the issue of security should be addressed in Nottinghamshire. Their proposal, at that meeting, was an armed association of defence. Initially the 4th Duke of Newcastle agreed in principle. Five days later, before the 4th Duke had had a chance to react, a police officer called to tell him that an inspector, sergeant and thirty men had arrived for duty in Mansfield. It became clear that this had been initiated by the Duke of Portland, concerned for his interests in Mansfield. The 4th Duke felt that he was having his authority as Lord Lieutenant usurped. He also now, on further reflection, had concerns that the expense of funding an armed association, which would fall on the population at large, who could ill afford it, could not be justified. The prime beneficiaries would be the wealthy property owners who, he felt, should pay for their own security. Neither the Duke of Portland nor Earl Manvers accepted that view.

This incident was illustrative of a change of national policy arising from lessons learnt during the Reform Bill riots: that there should be established permanent forces of constables to secure law and order. When fully implemented, which it was by the mid-1850s, it would usurp the yeomanries but, for the time being, the yeomanries retained an important role in aid of the civil power.

There was a meeting of the magistrates in Southwell in early March to discuss security when the 4th Duke opposed a nominee of the Whig Lord Chancellor to be appointed a magistrate on the grounds that the nominee in question was a Whig and a Dissenter. The 4th Duke attended the assizes in Nottingham in the middle of the month where security was also considered. In addition, he saw for himself on a visit to Mansfield, that the mood there was indeed hostile. As a result, on 22 March, he visited Nottingham again to meet with the magistrates and Colonel Sir Charles O'Donnell, the Brigade Major of the Northern (Military) District. There was concern about a public meeting of Chartists to be held on 26 March in Nottingham, as a result of which the 4th Duke, no doubt on advice, stood to both yeomanry regiments in a state of readiness. However the anticipated unrest did not occur.

In the meantime, the controversy over the nomination of the magistrate to whom the 4th Duke objected had not gone away. Indeed it had escalated to a dispute between the Lord Chancellor and the 4th Duke, culminating in the Lord Chancellor overruling the 4th Duke who then wrote an injudicious letter which he failed to withdraw when asked. He then received a letter, dated 30 April, from the Home Secretary, dismissing him as Lord Lieutenant of the county, quite an achievement in that day and age. It should be remembered, however, that this was a Whig administration and he was a right-wing Tory, implacably opposed to them, opposing the appointment of a magistrate selected by them, partly because the man was a Whig. The general opinion was that he had been in the wrong and should have bowed to the will of the Government, and the decision was not reversed even when he wrote a conciliatory letter. He was a man politically out of tune with the times. There is no doubt that the event shook him greatly.

He was succeeded as Lord Lieutenant by the then current Earl of Scarbrough, who resided at Rufford Abbey. This was John Lumley-Savile, 8th Earl of Scarbrough, Black Jack's second and only surviving son, Black Jack himself having died aged seventy-four in 1835, in a fall from his horse, whilst hunting near Doncaster. The 8th Earl had been MP for Nottinghamshire from 1826 to 1832 and for North Nottinghamshire from 1832 to 1835. It will come as no surprise to learn that he was a Whig. What is more surprising, however, given the times, is that although he never married he had a longstanding relationship with one Agnes 'Lumley', 'A French lady, who had been banished somewhat roughly from the home of her husband in her native country', without the benefit of a divorce. It was an enduring relationship and she bore him four sons.

Meanwhile the Government had become so concerned at the situation relating to the Chartists that, on 2 May 1839, the Home Secretary wrote to the commanding officers of each yeomanry in the Northern District, which included Nottinghamshire:

Whitehall

2nd May 1839

Sir, I have the honour to inform you that I have authorised the Major General commanding the Northern District to call out the Yeomanry of the counties of his district whenever it shall appear necessary. I have therefore to request that you will prepare your Corps in readiness to assemble as quickly as possible in aid of the Civil Power in case of emergency.

I have the honour to be etc.,etc.,
J. RUSSELL

J. Russell was Lord John Russell, the Principal Secretary of State for the Home Department, or Home Secretary in current terminology, a kinsman of the Duke of Bedford.

Immediately in its wake came the following, again to each yeomanry commanding officer:

NOTTINGHAM

4th May 1839

Sir

I have received the authority of the Secretary of State to call out your Corps of Yeomanry should I find it necessary to do so; the Lord Lieutenant of the County will probably inform you that this power has been given to me. I therefore earnestly request of you to hold your Corps in such a state of preparation that you can at once turn out if a requisition should be made to that effect. You are, no doubt, well acquainted with the threats of the Chartists to turn out on the 6th instant, from which day forward I trust that the zeal of your Corps for Her Majesty's service and the protection of the civil authorities will cause them to be particularly alert.

I have the honour to be etc.,etc.,
CHAS. NAPIER
Major General

The Major General was the officer commanding the Northern District, whose headquarters were in Manchester. This was the first sign of the development of another trend, namely the gradual erosion of the military powers of the Lords Lieutenant. It would be wrong to assume that the command of the Northern District could not attract a top-flight officer. General Napier already had an extremely distinguished record in the Napoleonic Wars and was destined to enhance his reputation even further in India.

The Sherwood Rangers had gone to Camp on 2 May, this year in Newark, using for training a large field at Holme, belonging to one of the 4th Duke's tenants, and so these letters must have added a considerable edge to the training.

Surprisingly, all things considered, there was considerable popular sympathy for the Duke concerning his loss of the Lord Lieutenancy. Indeed, from this moment onwards he seems to have mellowed and begun to enjoy much better relations with his neighbours and the county at large. Nowhere did this manifest itself more clearly than with the yeomanries, who knew him better than most, in that he now found himself being dined by each in turn on successive nights, firstly by the South Notts Hussars at Bullwell on 9 May ('In the evening dined with Colonel Moore and the officers – could not leave the table till past 12 o'clock – after sitting at the inn drinking tea, I left Nottingham for Newark, where I did not arrive before 3(am) and went immediately to bed.')

Then the Sherwood Rangers entertained him in Newark on 10 May ('William [now the Captain of one of the Troops, almost certainly the Clumber Troop] ... came to my room at 8 o'clock before I was up and gave me an account of all going on there ... dined with the officers as yesterday, and had to make a speech'). He had been required to make a speech the previous night as well. The phrase 'all going on there' was an understatement for on that same day, 10 May, the last day of the Regiment's period of annual training, the following letter was received by Sir Thomas:

Manchester 9th May 1839

Sir

I am directed by Sir Charles Napier KCB commanding the troops in the Northern District to desire that you will march the Sherwood Rangers from Newark to Mansfield there to remain embodied until further orders. The Major General has communicated with the Lord-Lieutenant of the County.

I have the honour to be Sir
Your most humble and obedient servant

J M Hughes

The Officer Commanding
The Sherwood Rangers Corps of Yeomanry

To which Sir Thomas replied:

In compliance with the above orders I have the honour to advise you that I shall march the Sherwood Rangers to Mansfield tomorrow morning at eight o'clock at which place I shall arrive at one o'clock in the afternoon

I have the Honour to be
Sir
Your most obedient humble servant

Thomas Woollaston White

The Right Honourable
Her Majesty's Principal Secretary
Of State for the Home Department

Sir Thomas no doubt then busied himself with the mobilization of his Regiment to be implemented the next morning. As the 4th Duke's diary reveals, he saw no reason to cancel the already planned Officers' Mess Guest Night arranged for that night, and no doubt the other end-of-camp parties in the other messes and canteens set out in the local hotels and hostelries, also took place as planned. It must have been a memorable night for all who were there. The Regiment marched out for Mansfield twenty miles away, as ordered, at 8 o'clock the next morning with their Commanding Officer's charger once more caparisoned in the crimson velvet and gold trappings of the great Duke of Marlborough. Frankly, the poor old Chartists didn't stand a chance.

The following is an extract from an instruction from Colonel O'Donnell dated 1 May 1839 in the South Notts Hussars' archives. It must have been conveyed to the Sherwood Rangers in the same form and gives a clear description of how they would have been expected to carry out the coming task.

So much depends on the readiness with which Cavalry can be brought to act, that I cannot too earnestly impress upon you the necessity of causing every Yeoman to be at all times (especially at the present period) prepared to turn out fully equipped for the shortest notice.

With this view I would recommend that he be frequently visited at his house and in his quarters by the non-commission officer of the Squad or Troop, and occasionally by his officer for the purpose of ascertaining that his horse is effective, and that his arms, ammunition, accoutrements, and horse appointments, and clothing, are in good order, properly put up in a secure place, and ready and fit for immediate use.

Let every troop have a central point as an alarm post, or place of assembly; such post should have a communication with each other, and with Head Quarters, for the early circulation of Orders and when the Regiment is together appoint one general place of assembly for the whole.

When placed in Billets in a Town, endeavour to keep each Troop as much together as possible, by dividing the Town into as many compartments as there are separate bodies.

Have stated hours for Stable duties, and require the attention of officers and men.

On all occasions march with an advance and rearguard; at times flankers will be necessary.

From five to six miles the hour is the ordinary rate of marching; and walk, trot and dismount alternately. Be careful not to distress the horses, particularly at the moment when it is probable an increase of exertion may be required.

When precaution is necessary, do not pass a defile, wood or suspicious place without reconnoitring it; for this purpose you can dismount a few men with carbines ...

Whenever Yeomanry are called out in times of disquietude to assist the Civil Power, quell disturbances, or protect property, the Troop should muster with the greatest promptitude. ...

No detachment ought to go without an officer who should have in his possession the General Order and Confidential Instructions of the 27th March 1835, and a Magistrate should accompany the Troop.

Better not to load except under particular circumstances, and then firing should only be resorted to at the very last extremity. The sword is the weapon for the Cavalry.

When you attack, let it be with great determination, and present as broad a Front as the nature of the ground will allow, observing as the general and most essential rule to keep a portion of your force strictly in reserve; too small a party should not be risked when there is the slightest possibility of defeat.

In the event of any immediate disturbance in or near a station, keep at least one-third of the Detachment in readiness and bridled as a Picket at one of the principal Billets or some central position, send out patrols and if necessary cause a part of the whole force to remain saddled.

A feed of corn may be reserved for each horse and carried with every Detachment when called out, but a third or one half of the horses only should be unbridled and fed at a time.

The bridles, swords, firearms, accoutrements, etc., should always be taken up by the men to their rooms at night.

On arrival in Mansfield they took over their billets, a process rendered routine for the Regiment and the populace alike by practice over several years of past annual training periods in the town. They discovered that the national leadership of the Chartists was planning to stage an initial rising at Sutton-in-Ashfield with the intention that this would trigger a spontaneous rising all over the country. However, they found the Regiment, fully reflecting the courage and determination of its Commanding Officer, so much against them and so determined to annihilate them if they made any show at all, that they thought better of it.

The Regiment remained embodied, albeit, no doubt disappointed, in Mansfield for the next two weeks. They focused on gathering information about the Chartists and their intentions which information was incorporated in regular written reports to Lord Russell. He later wrote a letter of thanks to the Regiment in which he said that these reports, containing information about the intentions of the Chartists, 'were the most correct in every particular of any that had been supplied to the Home Office during the whole crisis'.

On 23 May, having now been deployed away from their families, homes, farms or other means of employment, for a total of three weeks, inclusive of camp, and after an impressive performance, they were relieved by two troops of the South Notts Hussars. Those troops in turn were stood down on 28 May without encountering trouble.

Although the unrest had been contained it had not been entirely extinguished since there is a record that the Troop Commander of the Mansfield Troop, Captain Need (later Welfitt), attended

a meeting of the magistrates acting in and for the county of Nottingham, held at the house of Mr. Charles Neal in Mansfield Woodhouse on the 12th day of August 1839.

The meeting was chaired by the Duke of Portland, and resolved

> that the magistrates, with the assistance of the police and military, if necessary, do disperse the present, or any future chartist meeting at Mansfield, Sutton in Ashfield, on the forest, or elsewhere (...)

also

> search the houses of certain suspect parties suspected to be members of an unlawful combination, and of having in their possession arms for unlawful purposes, and against whom informations have been already made and warrants granted, and that such search should be conducted by such body of the civil power as shall be sufficient for the purpose, aided by the military force, if necessary (...).

As a consequence houses were searched and arrests made but there is no record of the Mansfield Troop being called out to assist.

In other parts of the country, where, perhaps, less decisive measures were taken than had been the case at Sutton-in-Ashfield, greater unrest had broken out. One beneficial result of the rioting that did occur was that it was then possible to prosecute the leaders which meant that they were jailed, thus ensuring there were no further disturbances in the country until they were released in 1842.

There is no doubt that the Regiment's performance and the level of recognition of this by the Government increased significantly the respect and affection in which it was held locally. The extent of this can best be judged by the events which took place when next the Regiment assembled for its usual annual period of eight days' compulsory drill, which took place between 26 April and 4 May 1840 in Worksop.

Firstly, Sir Thomas had been promoted to lieutenant colonel on 29 April during camp, and secondly, had also appointed, as his second in command, Henry Pelham-Clinton, the Earl of Lincoln, the 4th Duke's eldest son and heir. His appointment was also with seniority from 29 April although he was not present on this occasion. He was known as Lincoln until he succeeded his father in the 1850s. Lincoln was twenty-eight years old, which was on the late side for commencing a military career. Clearly, however, this turned out to be no passing phase or mere sinecure because he continued to serve in the Regiment, despite a successful political career, taking command of it in succession to Sir Thomas White in 1853.

Naturally, given the 4th Duke's involvement, Lincoln had been educated at Eton and then gone on to Oxford where he was President of the Oxford Union in 1831. By 1840 he was Tory MP for South Nottinghamshire, having been elected in 1832, and had already served as a Lord of the Treasury between 1834 and 1835. He had married, in 1832, Susan Harriet Catherine, the only daughter of the 10th Duke of Hamilton. Despite producing a family, including an heir, it was not a happy marriage, ending, unusually for the times, in a divorce in 1850. So now the 4th Duke had two sons in the Regiment.

On 4 May, described as a beautifully fine spring day, there was the final parade, the culmination of the period of training. It took place in the impressive old Manor Park at Worksop on the Duke of Norfolk's estate.

Normally there is a standard routine for these final parades at the end of camp but, on this occasion it turned into one of the most special events in the new Regiment's history to that date. For a start it took place in the presence of an unusually distinguished gathering including: The Earl and Countess Manvers; Lady White and family; Lady Mary Bentinck and Miss Bentinck; The Revd W. and Lady Frances Bridgeman Simpson; The Hon. Sydney William Herbert Pierrepont, now aged fifteen and destined to be the 3rd Earl Manvers; Granville Harcourt Vernon Esq. MP; Captain Lumley at that stage the heir to the Rufford Estate; Henry Savile-Lumley (later Savile) (1820–1881). The last-named was the fourth son of the relationship of the 8th Earl of Scarbrough, the Lord Lieutenant, with his partner Agnes and he was therefore illegitimate and unable to inherit the Earldom which would revert on the death of his father to other Lumley kinsmen and back to Sandbeck. However, he later became his father's heir to the Savile family estates of Rufford and Thornhill. These estates were very significant, consisting of some 34,000 acres, of which 18,000 were in Nottinghamshire.

From the foregoing it will be seen that all four of the ducal families were involved in one form or another, plus, in Captain Lumley, the representative of the other major Nottinghamshire landowner. Those five families accounted for almost 130,000 acres of Nottinghamshire. In addition there was a large gathering of county families.

The Corps of three Troops having formed a square, the carriages containing the principal ladies were drawn within it and the horses removed. Firstly, the Sherwood Rangers were presented with kettle drums, an exceptionally generous gift, by the 4th Duke of Newcastle, showing his undoubted affection for the Regiment as never before. Unfortunately he could not be present in person. A contemporary account recorded

> The drummer on a beautiful grey horse with the drums was ordered to the front. Colonel Sir Thomas White spoke a few words and presented the kettle drums and then read a letter from the Duke of Newcastle. Three hearty cheers were then given by the regiment and the band struck up "God Save the Queen" the whole corps singing the first verse at the same time. A proposed letter of thanks to the Duke of Newcastle was then read aloud.

Next the presentation of Standards took place. Lady White, who was of course Mary Euphemia White, the wife of Sir Thomas, presented the first Standard. No doubt in accordance with custom, it had been made by her or at least under her close supervision. It was rifle green in colour and carried on one side the county badge, an oak tree, and on the other side the horn and lanyard insignia of the original infantry regiment raised by Sir Thomas's father, and now to be adopted by the Regiment, under which was inscribed the motto 'Loyal until Death'. This was the White family motto, and from that day forward became that of the Regiment as well. It became the custom for it to be carried by the Mansfield Troop. The Standard remains a treasured possession of the Regiment. The account of the occasion stated:

> Lady Woollaston White stood up with dignity in her carriage and addressed the troops in a calm and clear voice: 'Gentlemen of the Sherwood Rangers, in presenting you with this Standard I have great pleasure in thanking you for the uniform attention and consideration displayed by you towards your commanding officer. It is also a matter of congratulation the numerous testimonials received by the Corps, not only from those high in office, but also from private individuals, evince how greatly your conduct and services were appreciated when last year you were called upon to repress the expected disturbances. Should that period ever unfortunately again arrive I am confident you will not be wanting; and in rallying under this standard you will never forget its motto 'Loyal until Death'.'
>
> Sir Thomas White then galloped up to her Ladyship and, receiving the standard from her hands, returned with it to the Regiment and, in their name, made an address of thanks. After this, whilst Sir Thomas held the Standard, a prayer was offered and the new Colours (sic) were consecrated by the Revd Charles Eyre (one suspects he was a kinsman of the late Colonel Eyre and possibly the inheritor of Rampton), addressing his words to Sir Thomas in the following terms:

> 'I consecrate this Standard in the name of God our Queen and our Country firmly persuaded that you Sir will forever do your duty so far as in you lies to your sovereign and the nation, and may the spirit give you grace to your duty to God himself.'

Next, the account reads;

> the right and centre Troops received standards from Granville Harcourt Vernon Esq., MP these standards having formerly belonged to the Nottinghamshire Yeomanry Cavalry commanded by the justly esteemed Colonel Eyre of Grove to whom they were presented in 1795 being the workmanship of and gift of Mrs Lumley-Savile and Lady Warren.

Granville Harcourt Vernon was, of course, the son of the late Colonel Eyre's second daughter.

An address from Lady Manvers, the eldest daughter of the late Colonel Eyre, was then read, after which she placed the two ancient standards from Mr Granville Harcourt Vernon in the hands of Sir Thomas White, who gave them to the standard bearers with a suitable address, and after a resolution for a letter of thanks had been passed, three cheers were given and the ceremony was concluded.

This was a magnificent gesture to, and endorsement of, the Regiment by the late Colonel's family. The Standards had, in the intervening years, been in safe keeping with Colonel Eyre. It is also interesting that these key possessions of the original Regiment should be re-presented to this successor to the original Regiment alone rather than to be shared with the excellent Southern regiment. Presumably, even at this stage, the seniority of the Newark Troop, through its unbroken service, was accepted by all as giving seniority to the Sherwood Rangers as between the two regiments. Thereafter the crimson Royal Standard was carried by the Newark Troop and the buff Provincial Standard by the Clumber Troop which was later renamed the Worksop Troop.

If there remains doubt as to the colour of the Regiment's uniform since 1828 it now becomes undisputed that from about this time that the Sherwood Rangers Yeomanry wore the rifle-green colours of the Woollaston White family's Sherwood Rangers for its uniform, but it is not known precisely when. It was not before Colonel Wildman took Command, and it had certainly happened several years before Sir Thomas relinquished command. Green was not a traditional colour for an English cavalry regiment, whether regular or yeomanry, and had never been a part of the Regiment's uniform at any earlier stage. Indeed, having initially been chosen, it had been specifically rejected as an option when the Regiment was raised in 1794, so it is hard to see the 4th Duke choosing it. Nor was it part of the tradition of any regiment in which Colonel Wildman had served, so it is unlikely that a decision on the colour of uniform taken by a combination of the 4th Duke and Colonel Wildman would have resulted in green, let alone rifle green, a strictly infantry colour. Given the introduction of other elements of the original Sherwood Rangers' regalia and traditions by Sir Thomas on this occasion, and the fact that green was not only the colour of the original Sherwood Rangers uniform, but of the Woollaston White family livery as well, it is highly likely that the decision was taken by Sir Thomas at some stage. It is unlikely that he would have sought to make such a fundamental change at the outset of his tour in command, because making new uniforms was expensive and, therefore, the existing stocks would have had to be allowed to wear out. Secondly, it is not something a newly-appointed commanding officer would have decided to implement until he had earned wide respect, based on a record built up over time. This he had now done. Therefore, since there was no mention of green being worn at the parade when such a distinctive colour never failed to draw mention in later years, these factors point towards the green being introduced shortly afterwards, since it was well established by the end of Sir Thomas's command.

When the green was introduced it was combined with gold lace and braiding. This is also unusual because, traditionally, gold lace signified regular cavalry and silver signified yeomanry. The combination of rifle green and gold satisfied the cavalry taste for bright combinations of colours. It was said to be 'together representing the colours of gorse in bloom'.

In 1841 the Regiment went to camp, as usual, at the beginning of May, this time commencing on 4 May at Newark. They used a field near Kelham Bridge as their training ground. There seem to have been several camps at this time which alternated between Newark and Worksop, since this was the fourth in that sequence. The 4th Duke's entry in his diary for 11 May reads:

Went this morning to Newark with my Daughters.

The Sherwood Rangers were reviewed in a field near Kelham Bridge – we arrived on the ground as the regiment was trotting past – all went off very well ... they cheered Lincoln who was out with them for the first time ... they then cheered me, which was perfectly unexpected. I acknowledged the compliment as well as I could. I afterwards dined at the mess which was a long business – and did not get back to Clumber until after 2 this morning.

He was clearly refreshing his memories of the stamina-sapping nature of soldiering with the yeomanry at annual camp, consisting, as it always has, with days filled with exacting training and nights filled with dinners, dances and other entertainments, famously and dubiously described as inducing 'simulated battle fatigue'. Of Lincoln he observed, 'He expresses himself much pleased with his new avocation of soldiering.'

All three Troops of the Regiment were again called out in 1841 for a day. The purpose is not recorded but was presumably connected to Chartism.

In 1843 the Regiment gathered in Retford for its annual training. At that time they used Grove Park, courtesy of the Vernon Harcourts, as their training ground, but would have been billeted in Retford. On 10 May 1843 the 4th Duke's diary records:

We all went this morning to be present at the inspection of the Sherwood Rangers in Grove Park - it really is a pretty sight – the ground being handsome, the day fair and a large concourse of people – probably 2000.

Also in 1843 the 4th Duke bought the Worksop Manor estate, the location of the parade in 1840, from the Duke of Norfolk who was selling to raise money. The 4th Duke bought it for the timber. He did not want the house and when he was unable to sell it he reluctantly had to demolish it to avoid the unaffordable cost of keeping it in repair. He paid £380,000, a huge premium. All but a small part was resold in

1890 for only £106,020 during a period when the price of agricultural land was high. This event ended the connection of the Dukes of Norfolk with Nottinghamshire. Given the poor relationship enjoyed between the two men, it is unlikely that the Duke of Norfolk would have sold to the 4th Duke, which is why the 4th Duke took the precaution of buying through an intermediary. Maybe the Duke of Norfolk did know, which is why he held out for such a high price.

The Regiment was invited to parade in Nottingham on 4 December 1843 when Queen Victoria, accompanied by Prince Albert and their State attendants, passed through on their way from Chatsworth House, the home of the Duke of Devonshire, to Belvoir Castle, the home of the Duke of Rutland. As the railway line to Nottingham was a terminus, they had to make the remainder of their journey by road. The Regiment's task was only to line the route and not to provide an escort. Also involved were the South Notts Hussars, the Inniskilling Dragoons and a detachment of the 64th of Foot. It is lucky that all that was called for was a route lining because the invitation was only issued to the units involved on 29 November, leaving precious little time, particularly for the Regiment to assemble, rehearse and march the twenty-to-twenty-five-odd miles to Nottingham. The 4th Duke noted in his diary that 'they are all in a commotion in consequence – it's a long way for them to go, and will I fear be found a great inconvenience for both time personal attention and expense.' Of course they duly went and were honoured to do so, moving out on 3 December. The Nottingham Journal reported the ground was 'admirably kept' by the two regiments and that there was a reception for them arranged afterwards by the Corporation in the Market Place to provide an opportunity to drink the Queen's health out of the capacious cup of the Corporation. Whilst the South Notts Hussars enjoyed a dinner that night in further celebration, the Sherwood Rangers faced the long march home.

Two other members of the White family served in the Regiment, both interesting for different reasons. The first was Sir Thomas's younger brother Taylor White (1805–1853), mentioned earlier. He served in the Clumber Troop between 1838 and 1844. In 1841 he took Holy Orders and became the Regiment's chaplain, serving in that capacity until his sudden and unexpected death in 1853; presumably he had inherited his father's heart. The second was Sir Thomas's son and heir, Thomas Woollaston White (1828–1909), who became the 3rd Baronet in 1882. He joined as a cornet in 1844 prior to a career in the 16th Light Dragoons, which he commenced in 1847. Retiring from the Sherwood Rangers in 1848, he had a long career in the 16th Light Dragoons, at least until 1870 when he retired as a brevet lieutenant colonel. He never saw action, but was nearly killed in 1867 by an enraged bull elephant he was hunting with a friend in India, which he shot but only wounded. The elephant charged and trapped them and attempted to trample and kneel on them and tusk them. In that situation death is nigh on certain but, for some reason, although both received multiple injuries, the elephant, perhaps due to the severity of its own wounds, was unable to finish the job.

Throughout this period, despite the fact that Britain was relatively free of the anarchy which afflicted many of its continental neighbours, Sir Thomas's orders were to keep the Regiment in a state readiness such that it was in a position to deal with any emergencies. To show how prudent were the foregoing precautions, in the spring of 1848 the Chartists made what turned out to be their final throw. Once again, Nottingham was at the centre of events since the 'arch agitator of the day', as he was styled, was one Feargus O'Connor, who was a member for the borough of Nottingham. The Chartists' plan once more was to establish an English republic. Given the general political stability that now prevailed and the overall growing prosperity, despite an economic setback at that precise time, it was even less likely to succeed than previously. The Chartists named 12 April as the day. In London, 200,000 Chartists rallied and marched with a petition to the Houses of Parliament. Similar rallies were planned in the county and, as a result, the Newark Troop spent several weeks maintaining order in Mansfield. This was sufficient to take the steam out of the situation and that was the end of Chartism as such.

The officer establishment of the Regiment at the end of the 1840s was as follows:

Rank	Name	Date Gazetted	Remarks
Lieutenant Colonel	Sir Thomas Wollaston White, Bart Wallingwells Park	29/4/40	JP
Major	Henry Pelham Pelham-Clinton, Earl of Lincoln	29/4/40	JP and future CO
Captains	William Samuel Welfitt	3/11/35	JP and served 17th Lancers and future CO
	Lord William Pelham-Clinton	11/4/38	
	John Manners-Sutton	3/2/44	JP Kelham Hall and future CO
Lieutenants	G. E. Harcourt Vernon	8/12/41	JP
	Hon. Sydney William Herbert Pierrepont	31/10/44	The future 3rd Earl Manvers

	William Leigh Mellish	18/4/48	Served in the Rifle Brigade as a Captain
	Jonathan Alderson	11/4/38	Served in 43rd Regiment of Foot
	Jonathan Hardcastle	31/1/45	
Cornets	John Vessey Machin	31/10/44	He was the son of Henry Machin. He was commissioned in 1851
	James Thorpe	4/6/47	
	Charles Thorold	14/4/48	
Surgeon	James P. Lacey	25/9/29	

It is interesting that G. E. Harcourt Vernon was of course the heir to the Headon and Grove estates. He was destined to command the Clumber Troop and have a reasonably successful political career as private secretary to Lord Lincoln when he was in government in the 1850s. He died prematurely in 1861 and bequeathed all his inheritance to his widow whom he loved dearly. Instead of them reverting on her death to Granville's brother as expected, she remarried, had a son and bequeathed the estates to him instead. As a result, to the Vernon family's dismay and the surprise of North Nottinghamshire, the land was lost to the descendants of the Eyre family, which may explain the lack of records concerning the Regiment in relation to its early days.

In 1849 one Trooper Ironmonger of Newark joined the Newark Troop. His father had served before him, joining when the Regiment was raised and so, given that this was the beginning of a period of service which would last out the century, it represents a significant period of involvement by one family. When Ironmonger joined, the Squadron Sergeant Major was one Bishop who had served in the Carabineers and had a well-earned reputation as a martinet. At the inspection of clothing on Ironmonger joining a large box of 'regimentals' was unpacked in the yard of the Clinton Arms in Newark where the troop was assembled. This may pinpoint the moment when the uniform was changed to green. Ironmonger recounted that there was no being measured for suits in those days; the uniforms were served out with as near an approach to fitting as the eye would allow. To achieve that, it was necessary for those being fitted to divest themselves of their outer garments. Ironmonger initially made the acquaintance of Mr Bishop when his nether garments were revealed as a result and it became apparent to the eye of Mr Bishop that they were not worthy of the Regiment or the wearer. 'Pull 'em off! Pull 'em off!' said Mr Bishop, 'then come for another pair.' This order was given as if he expected Ironmonger to divest himself of the offending garments in front of the whole Troop and other onlookers. Fortunately, an NCO intervened and Ironmonger was allowed to return to his home nearby and make the necessary adjustments.

Ironmonger made the comment that the uniform at the time was different to that of a Hussar, as worn for most of the remainder of the century. He said the jackets were of green embroidered with gold (no difference there). White trousers were worn on Sundays and on foot parade. The shako was of a large old-fashioned pattern, looking like painted tin. In his later years, Ironmonger told stories of the lighter side of yeomanry soldiering at this time and several of the anecdotes in this account are attributed to him.

In 1852, Sir Thomas Woollaston White, after a family connection with the Sherwood Rangers stretching for the whole of the sixty-three years of the Regiment's existence, relinquished command. To complete his story, in 1846, on the death of his former guardian Henry Gally Knight MP, he inherited from him over 3,000 acres adjacent to Wallingwells, covering the villages of Woodsetts, Gildingwells and Letwell, which he typically looked after in an exemplary fashion, notably rebuilding Letwell Church within eighteen months, after it had been destroyed by fire. Less happily in 1869 he had an accident in his carriage, when his horse bolted in Worksop, and lost a leg above the knee. His life was despaired for but, equally typically, he battled his way through to a complete recovery, or as complete as the circumstances permitted, and lived another twelve years. He had been an outstanding Commanding Officer.

Chapter 19

Loss of Role 1851

Sir Thomas White handed over command to the 5th Duke of Newcastle who had succeeded to the Dukedom on the death of his father on 12 January 1851. By that time Lord William had also died (in 1850). The 5th Duke was destined to command till his death in 1864 at the early age of fifty-two. The new commanding officer celebrated his appointment in a memorable way, entertaining the whole Regiment to dinner in a marquee erected at Clumber adjacent to the house on the evening of the final inspection and parade at the end of his first camp in command, which had been held at Worksop in May 1853.

It must have been quite some party. Sir Thomas White later reported that:

> When I looked through my window the following morning it appeared like a field of battle, for there were busbies lying in all directions, some men lying on the ground, others wandering as if lost, and horses loose and grazing.

From a historical point of view, this observation fixes the point at which the Regiment began to dress as hussars because, as has been noted, according to Ironmonger they were wearing shakos in 1849 but by 1853 they were wearing busbies. The change in dress was the opening initiative of the 5th Duke following his assumption of command. The colours remained rifle green with gold lacing.

In 1854 *The Times* of the May 12th contained the following report of the 5th Duke's second Camp which, as is recorded next, took place just as he was assuming the most challenging appointment of his political career:

> Yesterday the Sherwood Rangers, under the command of his Grace the Duke of Newcastle, completed their eight days' training at Newark. At noon the regiment was reviewed in a spacious field at Kelham, about two miles from Newark, and near the residence of Mr. Manners Sutton, one of the captains of the corps. The reviewing officer was Lord Francis Gordon, Colonel of the 1st Life Guards, who, accompanied by Sir Thomas Woollaston White, of Wallingwells, the late commanding officer of the Sherwood Rangers, and attended by an orderly of the regiment, arrived on the ground at 1 o'clock,

when the troops were put through about 20 movements. On the conclusion of the evolutions, they formed into a hollow square, and Major Welfitt proceeded to address them, telling them that they had gone through their duties to the entire satisfaction of the reviewing officer. Major Welfitt concluded by cordially thanking the men for the manner in which they had discharged their duties, and their admirable conduct in quarters. After several rounds of cheers had been given for the Queen, the Duke of Newcastle, Sir T.White, the reviewing officer, &c., the regiment formed into marching order, and returned to Newark.

By the time he took command, the 5th Duke had become a very considerable man in his own right. Since joining the Regiment in 1841 he had continued as MP for South Nottinghamshire until 1846 when he switched seats to the Falkirk Burgs, having fallen out politically with his father. He held the latter seat until he succeeded to the title. His politics were far more middle ground than his father's. He became a 'Peelite', a supporter of Peel who was a Tory, but on the left wing of the party, and who eventually combined with the Whigs and others to form both a Government and later the Liberal party. As can be imagined, he increasingly incurred the political displeasure of his father as a result, but not of his peers. He was appointed First Commissioner of Woods and Forests in 1841, an office he held till 1846, became a Privy Councillor in 1841. For some months in 1846 he was appointed to the Cabinet as Chief Secretary to the Viceroy of Ireland and was back in the Cabinet as Secretary of State for the Colonies from 1852 to 1854.

Much the most interesting appointment that he held was that of Secretary of State for War from 1854 to 1855. There can be few regiments that have been commanded by a serving Cabinet Minister, and even fewer, if any, that have been commanded by a serving Secretary of State for War, whilst, what is more, he had the conduct of a major campaign, the Crimean War. By the time he took office in 1854, the sequence of events which would result in the Crimean War being remembered primarily – despite its operational achievements – for its serious shortcomings, political, diplomatic, military and administrative, were already irrevocably in train.

The reasons for the conflict – the imperialistic ambitions of Russia to take advantage of the weakening Turkish Ottoman Empire by seizing its Danubian lands, Constantinople and the Black Sea – first emerged in the 1840s. The growing strength of Russia also threatened England's interests in India, and that gave England common cause with Turkey. This alliance was, in addition, supported by France for her own reasons. 1852 found the situation centring on a demand by Russia of Turkey that there be a general protectorate formed over Christians in the Turkish Empire whose position, the Russians claimed, was being threatened by Muslims.

It is said that the dispute, if correctly handled, could have been resolved diplomatically. However, this did not happen because of the failure of various parties, including the British Government, to react decisively when they had the opportunity. In addition, Turkey, knowing it had the support of Britain and France, became emboldened enough to reject Russia's proposal out of hand, rather than negotiate. Matters came to a head on 4 October 1853 when the Turks declared war on, and then attacked the Russians. Despite further diplomacy, which came too late and was too little, the Russians responded by attacking the Turkish fleet in the Black Sea. War was now inevitable. The aim of the Turks and their allies' campaign was to capture Sebastopol on the Crimean peninsula as a base from which to launch a campaign against Russia. This plan was flawed because the town turned out to be wholly inadequate for that purpose. It was at this point that the 5th Duke took the poisoned chalice of office.

Had the British Army been in good shape, all might still have been well. However, in half a century of peace since the end of the Napoleonic Wars, and despite the Duke of Wellington's continuing role as Commander in Chief 'his practices in the hands of lesser men had broken down', (as Churchill put it). This meant that the Army had failed to modernize; had succumbed to bad administration; and had failed to pay attention to the quality and appropriateness of the equipment with which it was being supplied. Worst of all, it had done little to ensure that the quality of the leadership had been sustained to the exacting standards required. Unfortunately, none of this was realized at the time and only came apparent as the campaign unfolded in circumstances calculated to expose any weaknesses. These were that it was being conducted on relatively distant shores, against a well-motivated enemy and, at times, in bitter climatic conditions.

The presence of a war correspondent for the Times, the first in history, ensured that the inability of the leadership on the ground to handle these problems was reported home. When these failures – the lack of the right clothing, and rations, and the rapid and unnecessary spread of disease and death through inadequate man management, leading inevitably to the unnecessary suffering of soldiers – were fully reported back, it was to an initially incredulous, then angry, public. The failure in 1855 of the first attack on Sebastopol left the 5th Duke with no choice but to resign. In 1856 Sebastopol at last fell, but its fall had by then been preceded by that of the whole British Government.

In helping to achieve the taking of Sebastopol, the British expeditionary force produced some of its finest feats of arms, in particular by the infantry at Alma and Inkerman, and by the cavalry through the charge of the Heavy Brigade at Balaclava. However the campaign would be forever remembered for two other events. Florence Nightingale and the light she shined on the plight of those wounded in combat or who caught contagious diseases because of poor hygiene in the field, and the ill-fated charge of the Light Brigade, misdirected by ambiguous orders to charge the wrong gun position. The Brigade itself cannot be blamed and are entitled therefore to claim it for themselves as one of the finest cavalry actions ever but, as Marshal Pierre Bosquet said, 'C'est magnifique mais n'est pas le guerre', not the last time in this history that those words will come to mind. One bonus was that the Staff College was founded as a result, and ever since Army officers have tended to be more precise, pedantic even, when expressing themselves in writing.

History has rightly not blamed the 5th Duke for the shortcomings of the campaign in the Crimea. He was nevertheless known thereafter as the Crimean Duke. The honourable way he accepted his fate and resigned led to his return, in due course, for his final ministerial appointment as Secretary of State for the Colonies once more, which he held from 1859 until his death.

Among many other positions he was appointed Lord Lieutenant for Nottinghamshire in 1857, and a Knight of the Garter in 1860, the third member of the Regiment, to that point, to be so honoured. He was described by Greville as 'a sensible man enough, but rather priggish and solemn, with very little elasticity in him' while Godwin Smith wrote:

He was not a great statesman, perhaps he was not even a very great administrator, for though he was a good man of business and devoted to work, he wore himself out with details which he ought to have left to his subordinates. ... he was a thoroughly upright, high-minded, and patriotic gentleman, who kept his soul above his rank, and devoted himself to the service of the State; while the fortitude with which he bore accumulated misfortune and torturing disease would have touched any heart.

According to Lord Selborne

He had a grave dignified presence; his reputation for integrity and honour stood high; and those who knew him best held him in great esteem. He was not, however, a good judge of men; his ambition was in excess of his powers.

These appraisals seem a little on the churlish side but they are expressed relative to the stage on which he played, and, by any normal standard, he had a fine career in which, impressively, his title seemed to play little part. The appraisal does not seem to reflect the nature of his command of the Sherwood Rangers, which seems to have been something he enjoyed enormously as a contrast to his huge responsibilities.

There was a story told by Arthur Lineker, who was a drummer boy in the Band, and of whom there is a photograph in the full-dress uniform of the regiment dated 1857. It is the earliest known photograph of someone in the Regiment and he served for many years. The story was of an incident during the 5th Duke's command which is illustrative of this. In 1857 the Regiment was at camp in Newark. The day dawned for the final parade. The Regiment was in position complete with the band. The moment came for the band to strike up with '*God Save the Queen*' but nothing came out. It was later said that the instruments, not the men of course, had been rendered inoperative 'because they had been wetted by beer or something else'. Make of that what you will.

The Duke instructed his orderly, one Wells, to approach the Band to tell them to 'play up!' This they did but it was not one of their finer performances. At the end of the Review the Duke said to the Regiment:

> I hope when you farmers return home you will find your pastures and meadows greener than your jackets and your cornfields at harvest time yellower than your braiding and may your wives, sisters and sweethearts be more harmonious than your Band!

The Clumber Troop under command of Captain John Vessey Machin, only recently promoted to command the Troop, was called out on Wednesday 16 October 1861 to provide an escort for the Prince of Wales from Retford station to Clumber Park on his visiting the 5th Duke and again on Saturday 19 October from Clumber to Worksop Station on his departure. The Orders read:

> The members of the Clumber Troop are requested to meet at the Normanton Inn on Tuesday next the 15th October at 1 o'clock in <u>Review Order</u>.
>
> The Captain is particularly anxious that every Member should turn out with the best horse he can possibly ride and to have his Kit in the best possible order.
>
> In the event of an escort being accepted one will be chosen from the Members of the Troop then Assembled for duty the next day.

J Vessey Machin
Captain

The Prince of Wales was, of course, Albert Edward (1841–1910) the eldest son of Queen Victoria and her husband Prince Albert who eventually succeeded her in 1901 to become Edward VII. At the time of the visit he was nineteen years old.

Arrival of the Prince at Clumber.

The visit lasted four days and consisted of a magnificent dinner party at Clumber on the Wednesday following his arrival from Cambridge where he was studying; an inspection of some features of the Duke's estate followed by a day's shooting on the Thursday; a visit to the Duke's collieries at Shireoaks and the laying of a foundation stone for the church the Duke was building in the village on the Friday; and the best part of a day's foxhunting on the Saturday from a meet of Mr Foley's hounds at Clumber before returning to Cambridge. The meet had been to open the foxhunting season locally which convention did not permit nationally before the beginning of November. The visit was hugely popular locally, attracting large and enthusiastic crowds whenever the Prince appeared in public.

At first sight this visit might be no more that an a pleasant duty undertaken by the 5th Duke who was doubtless very well acquainted with the Royal Family, designed to introduce the Prince of Wales to public duties. The Prince had already undertaken a major visit to Canada and the United States which had been judged a great personal success as well as successful in terms of the national interest. The 5th Duke, being Secretary of State for the Colonies, would doubtless have had a part to play in those arrangements and, therefore, would have become well known to the Prince, and so would have been an obvious choice to continue the educational process started by the visit to North America. However, given its content, which had clearly been put together with some care, the visit seems to have been much more about entertaining the Prince than training him in public duties. If so, given that the 5th Duke was a busy man, that would imply that the visit was much more about the creation of some quality time together during which important matters could be discussed and, if so, both the visit and the subject or subjects for discussion would have been initiated by someone other than the 5th Duke, almost certainly Queen Victoria herself.

Indeed, there was no shortage of possible issues. Prince Albert was desperately ill and, at the time of the visit, the Queen was consumed by concern for his health – with good reason because he died that December. The need for the Prince to have access to wise counsel he could both respect and trust would be important in such an event. The Queen was also seeking to arrange for the Prince of Wales to meet and marry Princess Alexandra of Denmark, the eldest daughter of Prince Christian of Denmark. Indeed, in September 1861, the Prince of Wales had been sent to Germany, supposedly to watch military manoeuvres, but actually in order to engineer a meeting between him and the Princess, a meeting which had indeed taken place and had gone well. With the benefit of hindsight this was hardly surprising because meetings between the Prince and members of the opposite sex invariably did go well. This was good news, however, because the Queen and Prince Albert had already decided on the marriage which eventually took place in 1863. However, this trait of the Prince had given rise to a significant problem that was causing very considerable concern: while visiting units in Ireland, brother officers had inserted a young actress called Nellie Clifton into the Princes' tent. One gets the impression that she was not immediately de-inserted by the Prince. If such a liaison both continued and became public knowledge during the Prince's betrothal to Princess Alexandra the social and diplomatic repercussions would have been significant. Finally, one way of keeping matters on an even keel until the marriage had taken place was for the Prince to partake in further official visits abroad. Such a visit was indeed arranged and took place in 1862 and no doubt the 5th Duke would have had a part to play in the arrangements. Whatever the truth, and whether or not the Prince was in need of being saved from himself or others, all could rest easy whilst he was under the protection of Captain John Vessey Machin and the Clumber Troop of the Sherwood Rangers Yeomanry.

The 8th Earl of Scarbrough, whom the 5th Duke had succeeded as Lord Lieutenant, had died in 1856. His longstanding relationship with Agnes 'Lumley' had resulted in no fewer than five sons. Because they were illegitimate, the Earldom, and the main Scarbrough estates in Yorkshire, Lincolnshire and Durham, left that line of the family and passed, by conventional rules of inheritance, to Richard George Lumley, a descendant of the fifth and youngest son of the 4th Earl of Scarbrough. He became the 9th Earl. Born at Tickhill Castle (the Scarbrough's ancestral home before Sandbeck was built in the eighteenth century) in 1813, indicating that his inheritance of those estates had been anticipated, and died in 1884, once he succeeded he lived at Sandbeck, and had little to do with Nottinghamshire. It is of interest to note, however, that he commanded the 1st West Yorkshire Yeomanry Cavalry, 1853–59, the predecessor of the Queen's Own Yorkshire Dragoons, of which more later in this account.

If there is no mention of actions by the Sherwood Rangers during this period that is for the very good reason that they took part in nothing which could be so described. The creation of a permanent police force during the course of the previous decade had superseded the role of the yeomanry nationwide. Nothing could illustrate this change more clearly than the violent Bread Riots which, once again, had broken out in Nottingham, this time in September 1854. This was an occurrence which had always previously needed the intervention of, at least, the South Notts Hussars, and for which they were indeed stood to, but was now dealt with single-handedly by the police, including a mounted detachment of their own men.

After sixty years of almost continuous use in aid of the civil power, the role of the yeomanries of maintaining law and order had virtually ceased to exist, and henceforth they would only rarely be required. This change was reflected by the decision, taken in 1856, that the yeomanries should come under the direct command of Horse Guards, with a primary role thenceforth involving purely military tasks, and only a secondary liability for call out for civil insurrection. This would remain the situation until reviewed by the Haldane Reforms of 1908.

Chapter 20

The Place of the Military in Society 1851–99

The country now found itself entering a period of relative peace and prosperity, which lasted until the end of the century. One of the reasons for this was that it was now under a Liberal Government which pursued a perhaps more temperate foreign policy than the Conservatives had in the past or would have done in their place. By contrast, in the rest of the world there was considerable turbulence. For example, the unification of Germany under the Kaiser and Bismarck and their growing influence in Europe, was exercised in a distinctly bellicose way with Germany invading Denmark in 1864, Austria in 1866 and, finally, precipitating a major war with France in 1870. Meanwhile the American Civil War took place across the Atlantic, and an imperialistic Russia flexed its muscles in the Balkans and in the Middle East and Asia.

It was the Royal Navy that took the brunt of the role of projecting Great Britain's increasing power internationally during this period. The activities of the Army were mainly focused on policing the country's interests in the Empire. It did so by taking part in a number of expeditions and skirmishes, mostly in Africa, culminating in the so-called River War in the Sudan of 1884 to avenge the death of Gordon of Khartoum. These commitments did not engage all the available regular troops and certainly did not engage the yeomanry, which was kept no more than ticking over. The increasing competence with which the regular Army executed the tasks given to it in the last half of the nineteenth century, demonstrated a steady improvement in its leadership and training. This was as a consequence of the reforms of 1856 and the lessons learned in the Crimea, which over time would extend to embrace the yeomanries and the militias raised during this period as well.

As a result the yeomanries and militias would find themselves – as indicated earlier – no longer simply a local defence force but increasingly part of the Army's order of battle. This meant that they would begin to work ever closer to and on an increasingly integrated basis with the regular Army, a change that, for all the differences of culture between volunteers and the regular Army, was welcomed to a far greater extent than it was not. All could see that closer links for training and access to the regular Army's greater professionalism, military expertise and resources was a benefit. Despite these improvements in standards, peace breeds indifference concerning the military in the mind of the population, and with it a loss of respect. This problem was increased by the fact that large sections of the Army, particularly that part which was at home and in sight, seemed to be under-utilized. This was because much of the Army was garrisoned in local towns throughout the United Kingdom and Ireland, so was highly visible, and came to be viewed with quiet irreverence, not to mention wry amusement, by the population at large in Great Britain.

Robert Smith Surtees, the novelist and satirist, who was to rural Victorian England what Dickens was to the towns and cities, caught this mood in 'Young Tom Hall'. He unkindly mocked the locally garrisoned regular 'Heavysteed Dragoons' that their only tasks seemed to be the attendance at each and every meet of hounds and each and every social gathering of the county, which their ample commanding officer, Colonel Blunt, described as 'yet another stomach ache and earache'. The volunteer element, which included the yeomanries, did not escape attention either. They had been the main instrument of law and order, and the symbol of authority of the ruling and landed class for so long, that they were not much loved by the population at large either. Now they had lost this role they were not viewed with even the same grudging respect that they had received before.

On the other hand the general prosperity of the nation was reflected in the ever more elaborate uniforms which they adopted so that Surtees's depiction of his fictional local yeomanry regiment which he called the 'Hyacinth Hussars', and then commented that they dressed 'as Dragoons by day and Hussars by night', in which order of dress they 'annihilated the ladies of an evening', contained more than just a hint of truth. It cannot be denied that, as already indicated, in the second half of the century, the Sherwood Rangers were second to none when it came to uniforms. They were indeed dressed magnificently as hussars (albeit by both day and night) in their striking rifle green and gold. They can at least claim that they had been comparing their uniform to the colours of gorse in bloom long before Surtees thought of his botanical parody: who knows, they may have even inspired it.

The existence of public prejudices relating to the yeomanry were described by Lord Harris during a debate in the House of Lords on 14 July 1887 as 'sneering attacks made at various times and in various places against the Yeomanry by men who had probably had never given a day's voluntary service to the country'.

The yeomanries, along with all other volunteer units, had, as an additional burden, the fact that their basic admiration for the regular Army was not widely reciprocated. They tended to be despised as incompetent by the regular rank and file. This attitude has, to a lesser extent, continued to this day in certain circumstances and in certain quarters, and has been extended to embrace the whole of today's Territorial Army. It is a charge that is not without basis, but as a generalization it does not stand up. The reserves, in the main, consist of well-trained and talented individuals who have created, with the help of their permanent staff, some very competent units.

The reason that it is important to draw attention here to what is a cultural difference is that, sooner or later, the two elements have to work seamlessly together, in order to perform to their full potential. This they cannot do in such a prejudiced atmosphere. Although, if this problem is to be overcome, it must take an effort from both sides. Nevertheless, the main initiative has to come from the regular element, in particular its leadership under whom the TA comes.

Reference to both dress and competence was made by HRH the Duke of Cambridge, the Commander in Chief, on the occasion of his inspection of the South Notts Hussars on 22 May 1891. This is an extract from his speech on that occasion:

> Ladies and gentlemen, but particularly gentlemen, although the ladies have a good deal to do with it,* I have a strong opinion in favour of the Yeomanry. I have often stated that they are decried; why or wherefore I cannot imagine, except that I suppose they are a little smarter in their dress than their neighbours. It seems nowadays that anything that is smart in this respect is rather looked down upon. Well, I take the opposite view, and I think whatever is smart ought to be supported, and you may depend upon it that smartness in dress has a great deal to do with efficiency. When a Regiment is smart in appearance and dress it is generally a good Regiment and in good condition.

One of the problems for the yeomanry was that of a combination of profile and perception. The Yeomanry in general – and the Sherwood Rangers are no exception – share, with all others who soldier, whether full- or part-time, a deep interest in all matters military. They have often chosen to conceal this behind a veneer of what Keith Douglas was to describe as 'their famous unconcern'. As a result they have often been mistaken as being indifferent to these things, sometimes to their detriment. Cavalry units have a higher profile compared with other arms. Riding the sort of high quality, fit, strong horse that is suitable for cavalry work under the weight of a man plus all his equipment and accoutrements, in company with other similar high quality, fit, strong horses, one-handed, demands horsemanship of a high standard. Well-fed horses tend to

---

* Is this the first recorded observation in praise of TA wives? If so it is the first of countless well deserved such tributes that have followed, to which number this comment can be added!

'hot up' in company. Even with experienced cavalry units it is realized that, where horses are involved together, vigilance, skill and professionalism of a high order must attend every moment. Things can and do go wrong. This combustible element differentiates totally the problem of training cavalry from almost any other military training of that era. There was simply no room for compromise. Even with highly skilled horsemen, horses will always manage to be horses, and have the ability to suddenly do things which will find the best of riders wanting. There is almost no way someone can be allowed to try and pick things up as they go along. Only people who have been riding horses for many years, or have undergone a full intensive military twelve-week course from scratch, from which they have passed out after a rigorous test, are fit to be allowed to train alongside others as part of a cavalry unit.

Naturally Yeomanry regiments, unlike regular cavalry regiments, simply could not, in the time available, even begin to train someone to ride from scratch up to the necessary standard. However, the problem was that they were also being pressed to recruit up to their establishment, irrespective of the availability of suitable horsemen. The only available compromise was to put the weaker

UP WITH THE SHERWOODS RANGERS

Trooper Stubbles (who has been repeatedly reprimanded by an officer for riding in advance and breaking the line). – OI CAN'T 'ELP IT, SQUIRE, IT'S ALL TH' HOULD MARE. HIVER SIN' OI LENT 'ER TO A CHAP TO ROIDE AT TH' EASTER MUNOOVERS, SHE'S BIN THAT HOWDACIOUS AN' WALIANT OI CAN DO NOUGHT WITH 'ER. SHE WEANT PLOO, AN'S SHE KICKS T'PIECES IVERY BLOOMING CART SHE'S PUT TO; AN' NOO SHE WEANT DO SOLDIERING UNLESS SHE'S FUST. YER'LL 'AVE TO FOIND ME ANOTHER 'OSS BY REVIEW DAY OR ELSE MAK' A HOSSIFER ON ME.

horsemen on docile and un-athletic horses, which were within their limited capability to control. The problem with that was that such horses and their riders looked ridiculous alongside the others and were not capable of keeping up with the horses of the correct standard, as their troop was put through the various movements. Nor could they keep going for long enough, lacking the necessary stamina. That being the case, either the regiment had to train at paces which were unacceptably slow or the formations quickly began to lose dressing as they trained.

Again, unlike other units, who mostly trained out of sight, none of this could be concealed by a yeomanry regiment as it went through the high profile process of an annual camp based in the local towns. Therefore it, coupled with the high social profile of some of the participants who served in such regiments, tended to fuel an assumption which was widely held that the yeomanry were much less competent and efficient than other volunteer units, and as a result also became the butt of ill-concealed amusement in the public eye.

Against this general background what was the public perception and more importantly the perception of the chain of command of the Sherwood Rangers during this period? Each year the Regiment was inspected by the cavalry commander in the Northern District, and all of these inspections were reported in the papers. In terms of professional competence, the reports were universally complimentary in terms of dress turnout, quality of horseflesh and ability to carry out the various movements and drills. This may not be considered as definitive, however, because such commanders were suspected of writing soft reports and it is necessary to look for other indicators; those tend to corroborate those reports as they applied to the Regiment.

The first is the reference to the uniformly high quality of the horses on parade. Year after year, not just the inspecting officer but the press as well, stated that the standard of horses was well up to that of a regular cavalry regiment. This is not really surprising. The Sherwood Rangers was a renowned hunting regiment, consisting overwhelmingly of countrymen working with animals all day, and therefore their horses were exclusively high-class hunters. Even if a yeoman did not own such a horse himself, one would have been provided from the officers' own studs or those of the Regiment's supporters, or the various livery yards which hired out hunters during the season. If there were only high-class horses in the Regiment's horse lines at camp then, for the reasons already given, there were no poor horsemen or horses unable to keep in line and, therefore, the drills are likely to have been of the competent standard that the inspecting officer reported, and the Regiment would have looked right and not ridiculous.

Every single moment of the Regiment's camps in those years was reported in the press at great length. A frequent comment was that the reporter was surprised at the high standards he witnessed against the public perception of the quality of the yeomanry generally. One example of this is a report of a joint exercise between the Sherwood Rangers and the Yorkshire Dragoons appearing in a national newspaper in1890 which opened as follows:

> To say that the auxiliary forces are in a very much better sense of the word soldiers now-a-days than was ever the case before is but to insist upon a fact patent to everyone who knows anything about the subject. Even the Yeomanry, who in bygone days were looked upon as fit and proper objects for the ridicule of small jokists no longer deserve, assuming they ever did deserve, the good natured obloquy which it was attempted to cast upon them. Every regiment vies with every other regiment in emulating the regular troops and in making its members as nearly soldiers as it is possible for men, who are first of all civilians, to become. A proof of this is afforded by the operations at present in progress between the Sherwood Rangers ... and the Yorkshire Dragoons ...

Perhaps the most conclusive indicator of all was the way in which the Regiment drew enthusiastic crowds of up to a couple of thousand lining the streets in every town in which they camped just to witness the Regiment marching out of town in the morning and back later in the day. In addition, crowds of several thousand more would come to watch the inspections and tournaments. They would not have been so supportive of a regiment which could not perform.

Perhaps the comment of one of the inspecting officers that the Sherwood Rangers was a 'smart little regiment' about summed it up. Either way there was no doubting their great popularity in Nottinghamshire.

Chapter 21

The Dukes, the Chain of Command
and Precedence

The Dukeries Families

In the years that followed the great families of the Dukeries and the other leading families of the north of the county were, for the main part, even more closely involved in the Sherwood Rangers, or, in the case of the Earls Manvers, and the other leading families closer to Nottingham, with the South Notts Hussars, than ever before. Not only did members of the family themselves serve in one or other of the regiments, but they were also encouraging those who worked for them and their tenant farmers to serve as well. They provided land and other resources, particularly money for uniforms and saddles and other accoutrements for the horses, and, of course, lent horses, the government's funding of the yeomanry not stretching to anything like all these things. Therefore a summary of the families in relation to the latter part of the nineteenth century is provided at the end of this account.

Command and Chain of Command:

In 1874 there was a major symbolic change. This was the year that Palmerston's Liberals lost power to Disraeli's Conservatives for the ensuing six years. The change was that the Yeomanry Cavalry, Militia and other volunteer units were placed directly under the War Office instead of being, as hitherto, under the authority of the Lord Lieutenant, completing a trend of diminishing powers of the Lord Lieutenant, which, as has been seen, had been developing over the previous twenty years: first, the loss of operational command to Horse Guards, then the loss of overall responsibility to the politicians. The position of the Lord Lieutenant after these changes in the structure of the chain of command was that he still signed volunteer commissions and continued to retain a valuable and influential pastoral role in respect of the reserves which extends to this day.

The Lords Lieutenants:

The 1st Lord Belper had succeeded the 5th Duke of Newcastle on his death in 1864. He was Edward Strutt of Kingston Hall near Kegworth who had been a

Liberal MP between 1830 and 1856, latterly for Nottingham, holding Cabinet Office as Chancellor of the Duchy of Lancaster. Ennobled as Baron Belper of Belper, he held office as Lord Lieutenant until his death in 1880. The 2nd Baron Belper was a committed Territorial and became commanding officer of the South Notts Hussars.

William Beauclerk, the 10th Duke of St Albans succeeded the 1st Baron Belper. The Dukes of St Albans owe their position to being the descendants of Charles Beauclerk, 1st Duke of St Albans, the illegitimate son of Charles II and his mistress Nell Gwyn. If it can be said that some of the Dukes whose history is recounted here owe their ennoblement to their role in the making of Kings, it can equally be said that in the case of the Dukes of St Albans the opposite was true. At this time they lived at Bestwood Lodge in Nottingham. The family's loyalty so far as the yeomanry is concerned is with the South Notts Hussars.

Finally, the 6th Duke of Portland, took over in 1899.

Precedence:

The question of Precedence is of limited interest to all but the units concerned and all parade adjutants down the ages. That is to say, when two or more units are on parade together, who is entitled to the 'right of the line'? It originates from the system settled for deciding the position that each unit took in relation to the others when forming up for battle. During the turbulent times in which the yeomanries had served prior to 1860 this had not arisen, because they mostly worked alone. In the stable climate which existed in the second half of the century, it was bound to be addressed at last because times of peace spawn parades of more than one unit, and you cannot have such a parade if you do not know the order of precedence of those taking part.

Precedence within the same arm is set by the date of formation. The establishment of the order of precedence in the case of the yeomanries was complicated by the fact that few were familiar with either the relevant rules or their own history, or had given the matter a second thought. It was, therefore, eventually established as a base line that so far as yeomanry regiments were concerned, the start date could not be earlier than 5 March 1794, the date of William Pitt's announcement to the House of Commons of his intention to raise such units. It was then ascertained that between 5 March 1794 and 14 November 1796, which was the high-water mark of the yeomanries in terms of numbers of regiments, fifty-one yeomanries comprising 225 troops had been raised.

As has been seen, originally yeomanries were raised as individual troops and only incorporated loosely into regiments. When the matter fell to be addressed in 1884, regiments were asked for the date on which their longest continuously-established troop had been raised.

Careful research has been carried out since, more detailed and better informed than was possible at the time, and on that basis the Sherwood Rangers date of seniority is 9 August 1794, the date the resolution raising it was passed. Therefore, of the original fifty-one regiments, the Sherwood Rangers, raised as part of the Nottinghamshire Yeomanry, would have been listed thirty-seventh in seniority in 1796. That being the case, by 1884 the Regiment would have ranked seventh, because all the troops which formed part of those which were originally senior, but were no longer so, had disbanded at one time or another, whereas the Newark Troop, which was one of the original troops forming the Sherwood Rangers, had never disbanded.

Those ranking ahead of the Regiment in 1884, in order of precedence, on the basis of the later, most accurate, research were: Pembrokeshire Yeomanry; Royal Buckinghamshire Yeomanry; Royal First Devonshire Yeomanry; West Somersetshire Yeomanry; Royal Wiltshire Yeomanry; Warwickshire Yeomanry. However, that is not the order of precedence that was laid down in 1884, because a number of the regiments, that had in truth been raised earlier than the Sherwood Rangers, did not then know their correct date. In addition, detrimentally for the Regiment, the date on which the Sherwood Rangers was raised that was officially identified, was 15 August 1794, the date on which the officers were gazetted and six days later than the correct date. For these reasons, the order of precedence officially set on 27 January 1885, and which has endured ever since, is not as stated above but records that the Sherwood Rangers are, out of the thirty-nine regiments listed on that date, fourth in seniority after: Royal Wiltshire Yeomanry; Warwickshire Yeomanry and Yorkshire Hussars. The reason the Yorkshire Hussars had been placed above the Sherwood Rangers was that they had been raised on 13 August 1794, between the Sherwood Rangers' correct date of seniority and the date they were officially given.

Whilst the confusion worked overall to the Sherwood Rangers' advantage, it worked cruelly against the South Notts Hussars who, because they were listed fifteenth in 1885 instead of fourteenth as should have been the case, missed the cut by one when, after the First World War, only the fourteen senior yeomanries were retained with a cavalry role. They became Royal Horse Artillery instead, a very real bone of contention for them at the time. Who is to say that in the end this has not served both well because, had they remained cavalry, the two regiments would almost certainly have been amalgamated in subsequent reorganizations, and lost their individual titles. As it is, to date, both still exist. In any event, the South Notts Hussars has since enjoyed a fine reputation, and war record, as an RHA Regiment.

It is amusing that such important inter-tribal boasting rights are based on such shaky foundations.

Chapter 22

Commanding Officers and Regimental Personalities

The 5th Duke was succeeded in command of the Regiment on his death in 1864 by Major (later Colonel) S. W. Welfitt, of Langwith Lodge. He was born at Mansfield Woodhouse in 1806 and educated at Rugby and Oxford. Born with the surname Need, he changed it to Welfitt on accepting a legacy from his uncle. He joined the 17th Lancers in 1826, but left that regiment in 1835. A High Sheriff of Nottinghamshire and a Justice of the Peace for Nottinghamshire and Derbyshire, his politics were said to be those of an ardent conservative. When he left the 17th Lancers he joined the Sherwood Rangers, initially becoming the Captain of the Mansfield Troop. He was destined to command for fourteen years. He was popular and the following letter to John Machin, still commanding the Clumber Troop, gives an insight as to why that was:

LANGWITH LODGE
MANSFIELD

Monday

My Dear Machin,

I send you three Black Flies all of which I find good here. The large one is the one you asked for. The middle sized one I catch most fish with. We go to London tomorrow and shall be away near three weeks, and the weeds will all going to be out so that I will ask you to fish her when I return, but if you like to like to have a day at Warsop with the fly you are welcome. I can't stand the inaction.

And now I am most anxious to thank you for the manner in which you commanded your squadron, the very excellent troop you have got, it never was so good during the 27 years I have known the Sherwood Rangers. Long may you live to command it. <u>Thank you Thank you</u>. I hope your wife is getting on well and with our very kind regards.

Very sincerely
J. M. Welfitt

This letter was post-marked Mansfield on 16 May 1863 and then Worksop on 17 May.

He was also clearly a good commanding officer as well if the following letter from the officer who carried out the Annual Inspection of the Regiment in 1864 is a fair reflection:

York
23rd May 1864

Dear Major Welfitt

You must at the same time allow me to congratulate you on the performance of your regiment. I don't know when I was so well pleased. Everything was so well done, the pace was so good and the turn out of both men and Horses so superior to what I have been accustomed to at my other inspections of Yeomanry.

Believe me
Yours very Truly

Signed M.V. Dickson

Note the importance attributed to 'pace' in relation to the points made earlier. The officers who camped at Worksop in May 1868 were as follows:

Officers 1868

Sub unit	Name and Rank	Remarks
Commanding Officer	Lieutenant Colonel Welfitt	
Second in Command	Major Manners-Sutton	Kelham Hall (Future Commanding Officer)
Newark Troop	Captain Thorpe	Coddington Hall, Newark (Future Commanding Officer)
Right Troop	Lieutenant Burnaby	Langford Hall, Newark
Left Troop	Lieutenant Anderson	Lea, Gainsborough
serafile	Cornet Mason	Cuckney
Clumber Troop	Captain Lord Edward Pelham-Clinton	The 5th Duke of Newcastle's second son
Right Troop	Lieutenant Monckton	Serlby. Future 7th Viscount Galway and Commanding Officer
Left Troop	Lieutenant Broadhurst	Ashley Cottage, Worksop
serafile	Cornet J. Thornley	Clumber, Acting Adjutant
Mansfield Troop	Captain William Frederick Webb	Newstead Abbey. Purchased Newstead Abbey from Thomas Wildman
Right Troop	Lieutenant Robinson	Mansfield
Left Troop	Lieutenant C. Tylden Wright	The Woodlands, Shireoaks
serafile	Cornet Walters	Papplewick Hall
Regimental Surgeon	Dr Wright	Ollerton
Chaplain	Rev. M. Hoole	

Lord Edward's inclusion on the Regiment's strength must have been in the nature of a guest appearance because, after the mandatory spell at Eton in 1854, he had joined the Rifle Brigade as an Ensign and served in the Crimea, but after the fall of Sebastopol. He reached the rank of captain in 1857 and then spent five years in Canada, returning in 1865. He then seems to have decided to base himself in Nottinghamshire because he joined the Regiment. He also decided to develop a career in politics and was elected unopposed at the 1865 General Election as the Liberal Member of Parliament for North Nottinghamshire. However, he clearly had a change of heart because he did not seek re-election in 1868, nor did he appear to continue to serve in the Regiment beyond that date. It seems that he decided to resume his career in the Army since, in 1878, he attained the rank of lieutenant colonel. It is also apparent that he must have been posted to India because he retired from the Army in 1880 whilst stationed there. Lord Edward then started a new career as a courtier and was Groom-in-Waiting to Queen Victoria from 1881 to 1894, then Master of the Household from 1894 until her death. Reverting to a Groom-in-Waiting under King Edward VII, he remained in that post until his death in 1907, by which time his full title had become Lieutenant Colonel Lord Edward William Pelham-Clinton GCVO KCB. Captain William Frederick Webb (1829–1899) had come, with his family, to reside in Nottinghamshire having purchased Newstead Abbey from Thomas Wildman's widow in 1861. Intriguingly he had to outbid Queen Victoria to do so. He had previous military service in the 17th Lancers and had retired as a Captain and had, no doubt, been recruited by Colonel Welfitt to command the Mansfield Troop as a result. He was very wealthy, having inherited estates in Yorkshire, Lincolnshire and County Durham. Although he is regarded as a generous custodian of both Newstead Abbey itself and the legacy of Lord Byron, Captain Webb became a renowned big game hunter and spent much time in Africa with his close friends Captain Codrington and the famous explorer Dr Livingstone. His trophies are still displayed at Newstead. Dr Livingstone is said to have saved his life on one occasion and was a frequent visitor at Newstead. William Webb was High Sheriff of Nottinghamshire in 1865 and was also a magistrate.

In 1878, Colonel Welfitt resigned his command of the Sherwood Rangers, after forty-four years of service, including the fourteen whilst in command. It was said that in forty-four years with the Regiment he never missed a single training (Camp). His great popularity caused him to be appointed Honorary Colonel, the first official such appointment in the Regiment's history, although informally that position had been held by several others. He was a keen hunting man, and took the Mastership of the Rufford Hounds during the 1860s, and was much praised, at the time, for the skill with which he did so. It was said that at the age of eighty he was still hunting, and indeed took a prominent part in a hunt of over two hours' duration 'across some of the stiffest of the Rufford country'. A universally popular figure throughout the County as well as in the Regiment itself, he gave extensively to charity. He died in May 1889 whilst the Regiment was at Camp. Those who attended his funeral, quite apart from many who knew him from the Regiment, read like a who's who of Nottinghamshire. His previous connection with the 17th Lancers was the harbinger of a relationship between the 17th/21st Lancers, now the Queen's Royal Lancers, and the Sherwood Rangers to be forged a century or so later, which has proved one of the most effective between a regular regiment and an affiliated TA regiment, and a good example of how the two cultures can work together to excellent effect.

Colonel John Henry Manners-Sutton, (1822–1898) of Kelham Hall was the next to command. He was the son of the Reverend Frederick Manners-Sutton and Henrietta Barbara Lumley, the daughter of John 'Black Jack' Lumley, 7th Earl of Scarbrough, himself a man of the cloth. John Manners-Sutton's great grandfather on his father's side was Lord George Manners-Sutton, the son of the 3rd Duke of Rutland of Belvoir Castle and the younger brother of Lord Robert Manners-Sutton, who served in the Regiment's predecessor which was raised for and took part in the quelling of the Jacobite Rebellion of 1745. Their eldest brother was the Marquess of Granby, the famous soldier. The Manners-Sutton family owned extensive estates in East Nottinghamshire, mostly in the parishes of Kelham, Averham, Rolleston, Knapthorpe, North Clifton and Harby.

John Henry Manners-Sutton after a short tour, in turn handed over to Colonel James Thorpe, of Coddington Hall, a prominent figure in Newark and, for many years, Chairman of the Newark (County) Bench of Magistrates and who had four sons all of whom were soldiers and one of whom was destined to command the Regiment. He, likewise, served for only a short time, being succeeded in 1882 by the 7th Viscount Galway.

7th Viscount Galway

It turned out to be a good day for the Regiment when The Honourable George Edmund Milnes Monckton-Arundell, later 7th Viscount Galway, joined the Regiment as a cornet in 1865. He was responsible, a few years later, for re-raising the Retford Troop which, as will be explained, was at a time that was critical to the Regiment's survival. The troop became known as the Tally Ho! Troop, because the Honourable George was a very fine horseman and he had recruited much of the troop from amongst his hard-riding hunting friends. This was the start of his remarkable involvement with the Sherwood Rangers which would last for the next sixty-one years.

The Galways' seat was Serlby Hall, on the north Nottinghamshire/ Yorkshire border. They had 4,081 acres in the north of the county and another 2,762 in Yorkshire. Considering the great role the Galway family was about to play in the history of the Regiment, it is surprising that there was apparently little or no previous involvement by the family with the Regiment, or apparently any militia either, until his father joined the Clumber Yeomanry. In fact, the family had been heavily involved militarily in the early years of the century, the 4th Viscount raising a substantial infantry unit, but centred on their local town of Bawtry, which was across the border in Yorkshire and, therefore, part of the process in that county and not in Nottinghamshire.

This next table shows its structure:

Yorkshire 1804	
Bawtry	
Rank	*Name*
Major Cmnd.	Robert 4th Viscount Galway
Captains	Andrew Cyril Jordan
	John Mathew
	Robert Graham
Lieutenants	Thomas Dawson Biochis
	Thomas Whitesmith
	Robert Sharpe
	Markham Nicholson
Ensigns	Joseph Blythman
	Robert Allan
Captain/Adjutant	William Sampson
Surgeon	John Drewater

One reason why the unit was based on Bawtry is that, in 1803, the widowed Lord Galway had remarried one Mary Bridget the only daughter and heiress of Pemberton Milnes of Bawtry Hall.

The 7th Viscount was the great grandson of the 4th Viscount and the only son and heir of his father. Born on 18 November 1844, he was twenty-six when his involvement with the Regiment began. He was educated at Eton and Oxford and was the Conservative MP for North Nottinghamshire from 1872 to 1875 and succeeded to the title on the death of his father in 1876; he was created

Baron Monckton of Serlby in 1887. This was one of the Jubilee peerages. A Yeomanry ADC between 1897 and 1901, he was appointed CB in 1903. Importantly for the Regiment, he was married in 1879 to Vere, daughter of Ellis Gosling of Busbridge Hall, Surrey. She later was appointed Lady of Grace of the Order of St John of Jerusalem, but she was best known for her love for and commitment to the Regiment, particularly during the hard times of the Boer War, when casualties were, for a time, all too common.

Fourteen years before, in 1856, Viscount Galway's uncle Horace Manners Monckton had married Sir Thomas Woollaston White's daughter, Georgina, in Woodsetts Church, thus connecting two of the most influential Regimental families.

In 1882, Viscount Galway's great tour of twenty-one years in command began. This is a list of the officers who served under him:

Officer	Troop/Appointment	Dates
Lieutenant/Captain/Major H. F. Huntsman of West Retford Hall	Newark/Mansfield	[1882]–1893
Captain/Major C. Wright	Clumber/2i/c	[1882]–1889
Lieutenant W. Watson	Clumber	1882–1886
Lieutenant/Captain W. Hollins	Mansfield	1882–1896
Lieutenant/Captain Champion	Retford	[1882]–1883
Lieutenant Mason of Morton Hall	Retford	1882
Lieutenant/Captain/Major E. E. Vernon Harcourt of Grove Hall	Retford	[1882]–1885
Captain Belwood	Adjutant	[1882]–1884
Dr Wright	Doctor	1883
Captain H. Bromley	Newark	1883–1889, rejoined 1900
Lieutenant/Captain/Major Lord John Savile-Lumley of Rufford Abbey	Newark/Mansfield	1883–1899
Lieutenant F. Cavendish-Bentinck		1883–1887
Lieutenant/Captain Ellis D. Gosling	Retford	1883–1894
Mr Martyn	Surgeon Major	1884–1892
Mr Martin	Veterinary Surgeon	1884–1889
Mr H. Dawson	Saddler	1884
Captain F. H. Blacker	Adjutant/rejoined 1892 as a member of the Regiment/ Mansfield	1885–1889, 1892–1898
Lieutenant Bosville	Retford	1886–1890

Officer	Troop/Appointment	Dates
Lieutenant/Captain T. R. Starkey, Norwood Park, Southwell	Mansfield	1886–1896
Captain/Lieutenant Colonel Denison of Babworth Hall and from 1898 of Eaton Hall	Retford	1887–1900
Second Lieutenant Lord Francis Pelham-Clinton-Hope	Clumber	1887–1892
Second Lieutenant R. C. Bacon	Retford	1890–1896
Dr Tristram	Surgeon	1889
Lieutenant Hickman	Retford	1889
Second Lieutenant/Lieutenant/ Captain J. F. Laycock of Wiseton Hall	Newark	1890–1899
Captain J. S. Willet	Adjutant	1890–1894
Lieutenant Seely	Attached from the SNH 1892 as part of the Regiment	1890–1892
Lieutenant T. B. Boswell	Retford	
Captain C. B. Harvey	Adjutant	1895–1898
The Revd Sir Richard Fitzherbert	Chaplain	
Lieutenant/Captain M. S. Dawson	Clumber	1892–1899
Surgeon Lieutenant C. Fleming	Doctor	1894–1899
Lieutenant H. Bromley	Newark	1894
Second Lieutenant/Lieutenant H. C. Peacock	Newark	1894–1899
Lieutenant Fitzherbert (son of the Padre?)	Clumber	1895
Lieutenant H. H. Wilson	Newark	1897–1899
Lieutenant R. T. O. Sheriffe	Newark	1897–1899
Veterinary Surgeon Reynolds		1897
Captain J. P. Jeffcock	Adjutant	1897–1899
Lieutenant H. Thorpe, (son of James Thorpe?)	Newark	1899–1900
Captain A. E. Whitaker of Babworth Hall	Mansfield	1899
Lieutenant the Hon. George Vere Monckton-Arundell, Lord Galway's son and heir	Retford	1900
Captain Younger	Lincolnshire	1900

Captain H. Bromley was Sir Henry Bromley Bart of East Stoke, Newark.

Here is some additional background on some of those listed above:

Major Lord John Savile-Lumley of Rufford Abbey, was the 2nd Baron Savile of Rufford (1853–1931), the son of the Reverend Frederick Savile, vicar of Bilsthorpe and nephew of the 1st Baron Savile. He had inherited the title because, as mentioned in the synopsis at the end of this account, it had been specifically remaindered to him when granted to the 1st Baron. He was educated at Eton and, predictably, given that he had been chosen by the 1st Baron to be his heir, entered the Foreign Office in 1873 but retired in 1889. Well known in sporting circles, he owned racehorses and was a good shot and fisherman. He married twice, his first wife predeceasing him. Although his second marriage gave him a son and heir, it was not happy and ended in divorce.

Lieutenant F. Cavendish-Bentinck (William George Frederick Cavendish-Bentinck, 1856–1948) was the grandson of Captain Lord Frederick Bentinck, the youngest son of the 3rd Duke, who had commanded the Mansfield Troop in the 1820s. He went to Cambridge and was a Trustee of the British Museum, Secretary of the Royal Commission on Education between 1886 and 1888 and a Fellow of the Society of Antiquaries. He was the father of both the 8th and 9th (who was also the last) Dukes of Portland.

Lieutenant/Captain T. R. Starkey was a member of the Starkey family of Norwood Park Southwell, a family, who normally joined the South Notts Hussars.

Captain and Lieutenant Colonel Denison of Babworth Hall and later of Eaton Hall was the grandson of Captain John Denison of Ossington, who had raised the Newark Troop in 1794, and the nephew of John Evelyn Denison, who also served, and who, it will be remembered, became Speaker of the House of Commons. Before joining the Regiment, Colonel Denison had already had a career in the Regular Army where he rose to the rank of lieutenant colonel in the Royal Engineers, whom he served for twenty years with distinction in different parts of the Empire. He was destined to command the Sherwood Rangers.

Second Lieutenant Lord Francis Pelham-Clinton-Hope was the younger brother of the 7th Duke of Newcastle and, as mentioned in the synopsis at the end of this account, was destined to succeed him as the 8th Duke of Newcastle when the 7th Duke died without issue.

Captain J. F. Laycock of Wiseton Hall was a member of the Northumberland Laycock family whose home there was Gosforth Park, the present site of Newcastle racecourse and had made their fortune in coal mining. They then purchased the Wiseton Estate and moved south and were destined to retain a close and distinguished involvement with the Regiment for many years.

Lieutenant Seely was John Edward Bernard Seely who became 1st Baron Mottistone CB, CMG, DSO, PC, TD (1868–1947). He was also awarded the Order of the Crown (Belgium), the Croix de Guerre, and the Légion d'honneur, the highest possible French award. The Seelys are a leading

Nottinghamshire family who had, like the Laycocks, made their fortune in coal mining, but in Nottinghamshire and Derbyshire, rather than Northumberland. Their family regiment was the South Notts Hussars, in which members served with comparable distinction to the Laycocks' service in the Sherwood Rangers, culminating with both families losing sons in the Second World War whilst in command of their respective regiments.

Jack Seely was the son of Sir Charles Seely, 1st Baronet, and was educated at Harrow where he met Winston Churchill, who became a lifelong friend. His main home was on the Isle of Wight where his branch of the Seely family had been based on their estate for many years. His friendship with Charles Beauclerk, the heir to the 10th Duke of St Albans had brought him to Nottinghamshire. They travelled together during the 1890s. It was not a long stay, no more than two or three years. Why he chose the Regiment rather than the South Notts Hussars is not known. The latter may have been fully recruited. Whatever the reason it was clearly an inspired decision, for having done so his career never looked back (!)

By the time of the start of the Boer War in 1899 Jack Seely was back on the Isle of Wight. As a result he had joined the Hampshire Yeomanry, and then commanded its service squadron during the Boer War during which he was Mentioned in Despatches, awarded the Queen's South Africa Medal with four clasps, as well as the DSO. Whilst still in South Africa he was elected Member of Parliament for the Isle of Wight and became a politician. In a charismatic political career in which he worked closely with Churchill, he rose to be appointed Secretary of State for War in 1912, with a seat in the Cabinet, a post he held until 1914. He was the second member of the Regiment to have held that office.

He was active in preparing the Army for war with Germany. The mobility of the proposed expeditionary force and, in particular, the development of a flying corps (the origin of modern day Royal Air Force), were his special interests. According to The Times, these developments played a significant role in the victory during the First World War. He was forced to resign over the Curragh incident in Ireland in 1914. He immediately joined up and left for the front in France. Already a very experienced cavalry soldier, having commanded in South Africa, within a short time he was promoted to brigadier general and given command of the Canadian Cavalry Brigade in Flanders. In that position he, of course, found himself under command of the very generals whom he had himself commanded only weeks before. He, with his charger Warrior, together the subject of an iconic portrait by Sir Alfred Munnings, enhanced their mutual reputation for bravery in battle. Known as 'Galloper Jack' Seely, he saw a great deal of action whilst in command of the Canadian Brigade which he commanded for most of the war, in both mounted and dismounted actions and took part in a full charge by the Brigade. After being gassed in 1918, he returned to England as the only member of the Cabinet, besides Churchill, to have seen active service in the war, and resumed his political career.

During the remainder of his life he held many important appointments both in and out of politics. In one he was appointed Chairman of the National Savings Committee. According to The Times, 'in the Second World War the activities of the National Savings Committee were largely extended and became a vital part of the national war effort.' He continued to have an influential role in domestic politics and his influence was considerable in the formation of the all-party Government of Co-operation in 1931. Indeed he is attributed by Lloyd George, through his work in that area, as being the 'Father of National Government'. The Times called him a 'Gallant Figure in War and Politics' and F. E. Smith, 1st Earl of Birkenhead, wrote, 'In fields of great and critical danger he has constantly, over a long period of years displayed a cool valour which everybody in the world who knows the facts freely recognizes.' He was ennobled in 1933.

The Reverend Sir Richard Fitzherbert (1846–1906) was the Rector of Warsop for many years but his home, which had been in the family for centuries, was Tissington Hall in Derbyshire.

Major A. E. Whitaker was of the Whitaker family who had made their fortune trading in Marsala in Sicilly earlier in the century and Major Whitaker's elder brother had purchased the Hesley Hall Estate near Bawtry in Yorkshire while Major Whitaker then purchased Babworth 'for the excellent partridge shooting it offered', and, indeed, still does. Major Whitaker commenced his military career in the 3rd West Yorkshire Light Infantry (Doncaster Militia) and, in 1879, passed into the famous 'Fighting Fifth' (5th Northumberland Fusiliers), with which he served in India. He took part under, General Lord Roberts, in the Afghan War, including the affair at the Daronta Pass, and received a medal for his services. While serving in that regiment he won the prize as the best shot among the officers of the Army. He was a well-known sportsman, and took a keen interest in miniature rifle shooting and was the president of the Retford Association. His unique collection of British and foreign war medals and orders is believed to have been one of the two finest collections in the country. First appointed to the command of the Mansfield Squadron on 4 February 1899, he was destined to command the Regiment. He held the tent-pegging championship of the Regiment for many years.

In 1882 Quarter Master Bousfield retired after forty-one years of service in the Regiment. He was dined out in spectacular style, details of the gargantuan menu served on that occasion surviving to astonish anyone who cares to read it, a feat only to be attempted on an empty stomach. Not much anecdotal information remains of his service but a fine portrait of him is in the possession of the Regiment.

As has been mentioned, another person who played a key role in the Regiment at that time was Quartermaster Sergeant Ironmonger who joined the Sherwood Rangers as far back as 1849. He was made corporal in 1859, sergeant in 1861, quartermaster in 1882 and regimental sergeant major in 1887. He resigned in 1897 with permission to continue to wear the uniform. By then he had achieved the unique record of having attended fifty trainings, or annual camps and, hardly surprisingly in the circumstances, been adorned with the title of 'Father of the Regiment'. He was a plumber by trade, which he carried out in Newark for very many years and was also, for some years, a member of the Town Council who was, in politics, a staunch Conservative, taking an active and prominent part in electioneering matters for the last half of the nineteenth century.

Chapter 23

Reorganizations

There is nothing like a period of sustained peace for triggering military reorganizations, the popularity of which, with politicians and the General Staff at least, never seeming to wane, irrespective of the lack of hard evidence that reorganizations, in every case, achieve any discernible long-term benefit. This period, so far as the yeomanries and others were concerned, was no exception. There had already, in 1856, been the change in the role and command structure of the yeomanry, already described. Secondly, in 1859, there were disagreements between the British Government and France, which looked as though they could result in conflict. As a result the Government developed the Volunteer movement, so as to be able to raise new reserve units throughout the country. This was successfully accomplished, with many new units of all arms and services being created. Many non–yeomanry TA units trace their origins to this reform. It also resulted in yeomanry regiments being permitted to improve their own recruitment levels.

The next reorganization, which came in January 1870, was triggered by the Franco–Prussian War. Although there was not a fundamental threat of yeomanry regiments being used it was decided that they should be put on a more economic and efficient footing. As has been mentioned, since the Sherwood Rangers had been established as a separate regiment, it had almost always been a three-troop regiment, consisting of the Newark Troop, the Mansfield Troop and the Worksop Troop. It is worth making the point that the Worksop Troop was now the Troop which was most closely associated with, and strongly supported by, the Clumber Estate, whereas in the past it had been the Rufford Troop, which had changed its name to the Clumber Troop.

Immediately prior to the reorganization the establishment of the Regiment was:

1 Lieutenant Colonel
1 Major
1 Surgeon

3 Captains
6 Lieutenants
3 Cornets
1 Veterinary Surgeon
3 Quartermasters
1 Sergeant Major
15 Sergeants
21 Corporals
3 Trumpeters
200 Privates
259 Total

The strength of the Regiment was only five short of that number.

The key danger for the Sherwood Rangers, arising from that reorganization, was that it was decided that unless regiments with three troops raised a fourth, they would be broken up. It was to solve this problem and save the Regiment from disbandment that the then commanding officer, Colonel Welfitt, prevailed upon the future Lord Galway to re-raise the Retford Troop. This matter of the small size of the Regiment, coupled henceforth with a chronic inability to recruit up to its given establishment from time to time, were to be its Achilles heels for the rest of the century. It was seldom able to report an effective strength of much above 75 per cent of its establishment, not even numbers equal to those when there were only three troops. As a result henceforth the regiment continually courted disbandment.

The main cause of the change from the good recruiting climate until recently enjoyed was the severe and lasting agricultural depression that now struck rural communities. This caused many traditional members of the Yeomanry involved in agriculture to have difficulty in being able to continue to afford to keep a 'nag horse' which, of course, doubled as their charger, and which was also essential if they were to be able to travel the considerable distances to attend training. The adverse trend now established ensured that each year saw a diminution in the enrolled strength of most regiments and so the Regiment was not unique, just on the small side.

However, why the Sherwood Rangers was a small regiment is not clear. It was unlikely to be a shortage of horses. While it may have been that rural Nottinghamshire was incapable of sustaining a larger unit, it may also have been that high standards of horsemanship were not compromised for the reasons given earlier, and sufficient suitable recruits of that standard could not be found. Another reason could have been that the frequent changes in the command of the Troops, where the responsibility for recruiting primarily lay, provided insufficient continuity. Increasing training obligations may also have been too much for too many people. As will be seen, all these reasons seemed to contribute

Captain William Webb commanding the Mansfield Troop at his home, Newstead Abbey, reporting to the commanding officer Lt Col S. W. Welfitt, 1870.

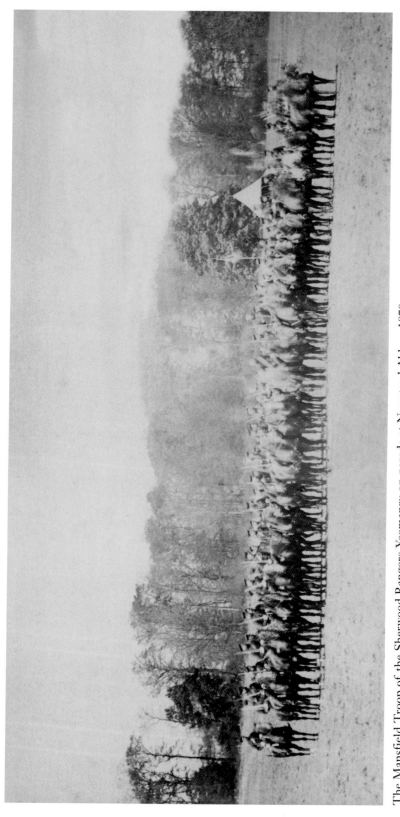

The Mansfield Troop of the Sherwood Rangers Yeomanry on parade at Newstead Abbey, 1870.

Lt Col John H Manners-Sutton, who commanded the Sherwood Rangers Yeomanry 1878–1881.

Quartermaster Bousfield served for forty-one years.

Lt Col James Thorpe commanded the Sherwood Rangers Yeomanry, 1881–1882.

Lt Col the 7th Viscount Galway, commanded the Sherwood Rangers Yeomanry, 1882–1903.

Lt Col the 7th Viscount Galway and the officers of the Sherwood Rangers Yeomanry (circa 1894). 'A smart little regiment.'

Brigadier J. E. B. Seely DSO TD on Warrior. He commenced his military career with the Sherwood Rangers. (Sir Alfred Munnings)

A field exercise between the Sherwood Rangers and the Yorkshire Dragoons, 1899.

The Sherwood Rangers defending the gates of Serlby against the Yorkshire Dragoons, 1899.

The C-in-C, Field Marshal the Viscount Wolseley, taking the salute from the Sherwood Rangers and the Yorkshire Dragoons, Serlby Park, 1899.

Holme Pierrepont Hall. (By kind permission of Bassetlaw Museum)

Rufford Abbey. (By kind permission of Bassetlaw Museum)

Welbeck Abbey.

Newstead Abbey.

Serlby Hall. (By kind permission of Bassetlaw Museum)

to the problem from time to time. Whatever the reason, the problem of recruiting the Territorial Army is as old as time, and few have found the complete answer.

The 1870 reorganization also went on to introduce the squadron system for the first time; so instead of there being one cornet per troop only one per squadron would be allowed. In addition to the size of each troop being specified, only one lieutenant per troop was to be allowed. This was not a popular reform because the traditional Troop system was much loved and preferred.

1871 found the German army advancing rapidly and victoriously through France and this shocked public opinion. The Government responded by commencing a series of long overdue reforms first identified as necessary during the Crimean War. These changes were resisted by the Commander in Chief, HRH the Duke of Cambridge, who was, of course, Queen Victoria's uncle. Edward Cardwell, Gladstone's Secretary of State for War, had to persuade a reluctant Queen Victoria to sign an Order in Council subordinating the Duke, under Cardwell's authority. That done, Cardwell and his successor Hugh Childers who between them abolished flogging, and the purchase of commissions, rearmed the infantry with the Martini-Henry rifle and rearranged the regimental system on county lines.

So far as the reserves were concerned, shorter terms of service for regular soldiers were introduced; this was intended to produce the material from which to build a more efficient reserve. Again this measure was of dubious benefit to the yeomanry since the last thing a former regular soldier usually wants to do on leaving the regular Army is to join a reserve unit. Alternatively, if he is simply placed on a list of reserves with no training obligation, 'skill fade' quickly reduces his usefulness. For these reasons, either way, only a limited number of ex-regular soldiers go on to give valuable service to the reserves. Those that do, however, are invaluable.

Among other measures introduced was a considerable tightening of arrangements relating to the yeomanries for training, which have already been mentioned, and which involved a significantly increased commitment. The reform sought to combine more frequent voluntary drills, stringent standards of musketry and lengthy drill courses for officers. Three mounted troop drills now had to be carried out by troops before and in addition to assembling for annual duty. The recruit training was also tightened up and recruits had to go through a course consisting of twelve drills and six squad drills (mounted and dismounted). Because many people could not meet these increased commitments, this contributed significantly to the recruiting problem.

This is a classic example of the kind of seemingly logical decisions imposed by the chain of command, that partly fail in their purpose because such decisions are not made with the benefit of the advice of those who understand volunteer service. A fine line exists, when seeking to achieve improved efficiency by stepping up training regimes, between achieving that aim, and ending with worse standards of readiness than existed before the new regime was introduced. In this case, although those who received the additional training were better prepared so far as individual skills were concerned, they were worse prepared in another way, because the loss of strength reduced the quality of collective training.

The changes clearly had a big impact on the Sherwood Rangers. In 1870, when only three troops, their recruited strength had been 245 against an establishment of 250. By 1883, with four troops, their recruited strength was 166 against an establishment of 226.

1876 was the year of the 'Great Mobilization Scheme', a plan designed to protect the nation from invasion, which given that it never rose above being solely paper-based, will have struggled to strike the planned for 'fear and awe' into our enemies. It formed the whole Army, including reserves, into seven army corps. Its drawback was that the units were not allocated to corps on a geographical basis and so were never able to train together. In addition, the formation staffs were not appointed, so that units never even came under direction. Nevertheless, this structure was recorded in the Army List for a number of years. For the record, the Sherwood Rangers were posted, with the South Notts Hussars, as divisional Troops to 2nd Division of VI Army Corps.

In 1893 there was a significant reorganization of the yeomanries. The first element was the abolition of the Troop system which was replaced by squadrons, each consisting of two troops, with the senior troop commander taking command of the squadron. But, hold on, had not the Troop system already been abolished in 1870? Why does it come as no surprise to learn that the attempt by the chain of command to abolish the popular Troop system back then had seemingly fallen prey to that old 'yeomanry deafness' problem? This time the reform was implemented as part of the 1893 reforms. It was clearly part of a plan to overcome the falling strength of the yeomanries as a result of a combination of the reorganization of 1871 and the continuing agricultural depression. It laid down that each squadron would have an establishment of 100 and, if it fell below a strength of seventy 'efficient' members, was liable to be disbanded. Regiments were to be formed into two-, three- and four-squadron regiments with uniform establishments of 218, 324 and 430 respectively. The Sherwood Rangers, being a four-troop regiment, became a two-squadron one with an establishment of 218. This involved a further threat to them because there was an additional directive that if a two-squadron regiment fell below 140 'efficients' it was liable to be broken up. The Regiment was in fact threatened with disbandment in 1889 on the grounds of its poor strength, but in 1890 it went to camp 200 strong amidst a strong sense that this had been a very close shave.

The new 1893 establishment of the Regiment was to be:

1 Lieutenant Colonel
1 Major
4 Captains

    4 Lieutenants
    2 Second Lieutenants
    1 Medical Officer
    1 Veterinary Officer
    4 Permanent Staff Sergeants
  17 Sergeants
  12 Corporals
    4 Trumpeters
167 Privates
218 Total

The Regiment managed to camp with comfortably more than the 140 minimum to avoid further risk of disbandment, but the smallness of the Regiment remained a problem for all that.

    The thirty-nine regiments of yeomanry were formed into eighteen brigades, sixteen of two regiments and two of three regiments, leaving the Pembrokeshire Yeomanry un-brigaded due to being allotted to the Milford Haven defences. The eighteen brigades were numbered one to eighteen, but three were also respectively named the Kent, Portsmouth and Devon Brigades. The Sherwood Rangers were brigaded with the South Notts Hussars in 17 Brigade. There is no record of the basis on which numbers were allocated.

    The brigades were to be commanded by the senior yeomanry commanding officer which was a popular decision. In Nottinghamshire it was Lord Belper (the 2nd Baron Belper), the highly regarded commanding officer of the South Notts Hussars, who took command. The brigade adjutant was Captain J. Slattern Willett of the King's Dragoon Guards, the adjutant to the Sherwood Rangers. On mobilization, 17 Brigade would become the Divisional Yeomanry Cavalry for 9th Division.

    One of the most resented parts of the reorganization was that it was used to cut the number of regular officers posted as adjutant from one per regiment to one per brigade. Lord Galway complained strongly about these reforms in the debate in the House of Lords in May 1892. Lord Brownlow, on behalf of the Government, explained that the real problem was the small size of the average yeomanry regiment. Of the thirty-nine existing regiments, sixteen were less than 200 in number, whereas the Government was looking for a yeomanry regiment to be about 400 in number. The idea of the Brigades was to effectively bring that about without the loss of cap badges. The Government was also concerned at the poor marksmanship to be found in the yeomanries. This was true because the yeomanries saw this skill as less important than those of conventional cavalry, and resented the emphasis that the introduction of a payment qualification based on ability to shoot placed on that skill against others. The background to this particular bone of contention will be explained next.

Chapter 24

Role and Equipment

When HRH the Duke of Cambridge inspected the South Notts Hussars on 22 May 1891 he raised the vexed issue of role: whether the yeomanry should remain cavalry or become mounted infantry.

> I know you hear about mounted infantry. Well it may be very good in its own way, but I like Yeomanry better. You have a very strong advocate in your favour. I know other views are taken but it is not necessary for me always to agree with them. I do not mind saying when I do differ, as I do differ absolutely, I say I am always happy to support the Yeomanry although I have the highest opinion of all other portions of Her Majesty's Service. There are different elements in the Army: I say take advantage of all of them, and do not discard one for another. There is room enough for them all, and as there is room for them all, why on earth should we get rid of one simply for the pleasure of change? I don't like changes. I always set my face against them. Changes are necessary in this world, alas! Still let them be prudent and advisable not rash and unnecessary.

The primary role of cavalry was under review because the Franco-German Wars of the 1870s had shown that modern weapons, particularly well-sited machine guns, had the beating of it by a considerable measure. Indeed, the First World War became entrenched in the way it did precisely because of this, and by the beginning of the Second World War the role formerly fulfilled by cavalry had passed to the tank.

In the case of the cavalry at the end of the nineteenth century, the whole debate concerning cavalry and mounted infantry was happening in a period when each was as obsolescent as the other. That said, up to and including the beginning of the 1940s, during the early years of the Second World War, cavalry continued, on occasions, successfully to deliver effective action, even against machine guns, when it was able to achieve tactical surprise. Indeed it could still do the same today, but there is no point given the existence of the protection and other advantages offered by armour.

It sometimes seems that over a century later this whole debate has passed over the heads of the dismounted infantry. Of course the nature of the role requires them to take to their feet at times without substituting the protection provided by armour with fortifications whilst in a defensive mode or close country when mobile, although, it seems that, all too often, they allow themselves to be deployed on their feet in battlefield situations where the ground is commanded by weapons against which they have no adequate defence.

Mounted infantry was the mere beginning of a long and continuing saga of roles that yeomanry regiments would be asked to perform, as their original role as cavalry gradually became less relevant and eventually ceased to exist altogether. There is virtually nothing that one yeomanry or another has not by now turned its hand to, and usually with considerable distinction. Although no one likes to be re-roled from a preferred arm or service, the reality is that these changes have always been a feature of yeomanry soldiering. In the end the only acceptable reaction, once the change is inevitable and justifiable, is to accept and make the very best of it. Tradition, culture and who is alongside you are far more important than role. That said conventional wisdom would claim the Sherwood Rangers have generally been 'lucky' in the roles assigned to them over the years, which have generally been, if not their preferred option, a palatable alternative.

Before seeking to explain the difference between cavalry and mounted infantry so that the controversy can be appreciated in this age of armoured and airborne warfare, it is worth seeking to justify why the debate has a place in the history of the Regiment. After all, although it called itself a cavalry regiment, it had, in almost a century, only ever been used as a form of police force. The reason is that, in the fifty years between the late 1890s and the early 1940s when the Sherwood Rangers lost their horses altogether, the game would change dramatically, and they would see action as both cavalry, mounted infantry and several nuances between the two in three separate wars. They were, in particular, destined to play a leading role, along with several other yeomanries, in perhaps the finest action ever by British cavalry – Allenby's advances against the Turks in Palestine from Gaza to Damascus and from Jerusalem to Aleppo in the First World War.

The main role of cavalry was to overwhelm an enemy position through the delivery of 'Shock Action', a phrase which means a combination of surprise, speed and sheer weight of arms. This was achieved by manoeuvre which caused cavalry to arrive unexpectedly, or earlier than expected or from an unexpected direction, the surprise thus generated un-nerving the enemy and degrading his ability to respond and will to fight. In each case, to create the full effect, it had to be done with sufficient weight of arms to overwhelm the enemy. The coup de grace was usually delivered en masse in close formation in a single strike using a pointed weapon, usually a lance or sabre with which to run the enemy through or cut him down. If the enemy was not routed, the formation then pulled up, turned round and charged back to finish him off. Once defeated, the enemy was pursued and cut down or captured. The charge was not easy to achieve and the return charge even less so and British cavalry, it must be said, does not have a great record in that department.

The subsidiary but equally valuable roles were reconnaissance, scouting, patrolling, providing mobile picquets, escorts and liaison.

The primary concern was that although the cavalry charge looks a fairly unskilled event it actually took a lot of training to bring off effectively and a level of skill – including the use of the sabre as well as in general horsemanship – which was thought to be beyond the capability of a yeomanry regiment to achieve. The argument in favour of yeomanry retaining the role was the capability to execute a charge since even a poorly executed one could carry a situation which would otherwise be lost for the want of being equipped as cavalry, that is to say with a lance or sabre. In addition, unless so equipped, a mounted formation would be unable itself even to attempt to resist a charge by cavalry, because rifle fire alone was insufficient, and being issued with either lance or sabre essential.

The concern with the role of mounted infantry was that the horses were used simply to transport the regiment onto the battlefield and then were taken to the rear whilst the unit went into action as infantry, which meant there was little need to learn cavalry tactics or to develop the all-round skills of horsemanship and horse-mastership that were the sine qua non of the cavalry. The Yeomanries were united in the view that their regiments should be equipped with both sabre and carbine so as to keep the option of being used as cavalry open to them.

One of the problems the regular chain of command has in assessing the competence of a volunteer – which is illustrated here – is that when assessing a regular soldier they only need to take into account what he has learned whilst being trained by the Army, because every skill he possesses is clearly documented on his record and is familiar to them. This is simply impossible to replicate in the case of a Territorial because the military record is only a lesser part of his whole experience. This means that more than half his total skills are ignored entirely when capability is assessed. As a result, in the present case, assessing whether a man could ride or handle a weapon such as a sabre they would ignore, for example, that he worked with horses for a living, or was used to using blades of all sorts, or hunted as a pastime which called for horsemanship of a high order. Although these skills are by no means fully comparable they are relevant to the assessment of what the volunteer is capable of achieving and should be identified and taken into account. This issue is as relevant today as it was then. It can also work the other way in that it can create an over ambitious assumption of capability in the volunteer.

Chapter 25

The Training Year

Superficially, life in the Sherwood Rangers continued after the changes in role of the 1850s just as it had evolved in the years following the end of the Napoleonic Wars. The key event in the training calendar was the eight-day 'Training' or Annual Camp, as it later became known, which, by the end of the nineteenth century, had increased to ten days. Due to the difficulty of gathering the Regiment, or even single troops together, caused by constraints imposed by travelling distances and time, this was the only real opportunity each year for collective training. The remainder of the year was devoted to recruiting, individual training and equipping the Regiment.

There were no drill halls, the Worksop Troop, for example, meeting at the Normanton Inn instead. There was also the all-year-round commitment of the agricultural seasons. Given the foregoing, it is easy to see how difficult it was for much more military training to have been attempted in the rural economy, of which most members were part and which was, in addition, in recession for much of this time. Despite these constraints on increased demands being imposed, from 1871 more frequent voluntary drills, stringent standards of musketry, and lengthy courses for officers were introduced. Recruit training was also tightened up and recruits from then on had to go through a course consisting of twelve drills and six squad drills (mounted and dismounted), a well-intentioned increase but not necessarily effective in the circumstances.

As mentioned the Regiment now dressed as hussars, and looked magnificent in busbies, brown bearskin for officers and black bearskin for other ranks, and rifle-green uniforms embellished with gold braid. They continued to dress as such for the rest of the century until khaki became universal.

If the year outside of Annual Camp provided insufficient time for training, the period of annual camp itself suffered in a different way, as it developed into the major challenge of compressing the whole of a year in the life of a regiment into a period of eight days, a feat which became a standard part of volunteer peacetime soldiering in the nineteenth and twentieth centuries. As has been described, since the concept of annual camps had been introduced in the 1820s, camps were based on one or other of the towns with which the Troops of the Regiment were associated from time to time, namely Newark, Mansfield, Worksop and Retford, and this had become the tradition. We have seen that the absentee troop during the mid years of the nineteenth century was the Retford Troop. The Regiment therefore camped at each of Newark, Mansfield and Worksop on a cyclical basis, with Retford included once the Retford Troop was re-established. The reason why the towns, rather than the great estates, were still being chosen as the base for camps, even though the role in relation to law and order had almost ceased was because yeomanry regiments did not have integrated logistical support, without which no Regiment can survive as a field unit. Everything they needed, however, was to be found in the towns which overcame the problem. The reason the towns remained the same ones originally chosen for operational reasons that were no longer relevant was the familiarity of each of the towns with the needs of the Regiment. Also, the Regiment needed a large open area of grass with good going and free from hazards such as rabbit holes, in order to practise their drills and carry out other training, and such areas were not easily found. When one had been established, it made sense to camp near it and use that, rather than try and find and clear a fresh one each year.

The area used when the Regiment camped at Newark was a large field just outside the town on the road to Kelham belonging to Colonel John Manners-Sutton. At Worksop it was an area called the Plain Piece, in the woods at the top of Sparken Hill, effectively in the grounds of Worksop Manor and regarded as the best of all their training grounds. This belonged to the Dukes of Newcastle. At Mansfield it was an area at Radmanthwaite and at Retford it was originally Grove Park, the home of the Vernon Harcourts, or Headon Park, the Eyre family's home, next door. When Colonel Denison took over command of the Retford Troop he made his home, Babworth Park, available. The Regiment would march out on horseback from the town to the training ground to train and march back each night.

Camp invariably took place in the first week of May, a tradition that was established in Colonel Wildman's tour of command. The reason this time of year was popular was that it fitted conveniently with the farming cycle, after spring drilling had been completed and before the start of haymaking. It was also after the end of the hunting season and before hunters had been finally roughed off and turned out. It became such a tradition it was continued years after the typical yeoman was more likely to be a coal miner than a farmer.

The programme for each camp was pretty much identical no matter which town was being used and was divided between intense military training and an almost equally intense social programme. The following is typical:

Day 1: The Troops would assemble in their home town. The 'host' Troop, having had responsibility for much of the arrangements for the camp, would now have the compensation of a short ride. The other Troops would take the day to march to camp, stopping for lunch at traditional locations, always wayside inns, in order to rest and water the horses. If modern-day yeomen have ever wondered where the talent and tradition of the yeomanry for blending roadside hostelries into their operational imperatives came from, they need wonder no more.

Sometimes the move to camp was, but only after the halt for lunch, turned into a training scheme in which the host Troop would send out patrols along the other Troops' likely lines of march to try and intercept and, unseen, report their progress back to RHQ. Meanwhile, the marching troops would similarly put out an advance screen to prevent the home troop from gathering the information it was seeking. As the role of yeomanry cavalry changed from aid to the civil power to a purely military one, battlefield reconnaissance became the premier role and, for those yeomanry regiments that remained part of either the cavalry or its successor, the Royal Armoured Corps, has remained such ever since, subject to slight variations. Only the equipment used has changed from horses to vehicles of one sort or another.

On arrival Troops took over the accommodation and stabling allocated to them, the officers occupying the best hotel, or a brother officer's house and everyone else fitting in round that.

Towards the end of the century the Troops with furthest to travel started to travel by train.

Day 2: The Regiment would mount Parade in the market square at 08.30, to achieve which involved Reveille at, at least, 05.30. They then marched out to the Training Ground for mounted Troop and regimental parade Drills. Cavalry movements require good horsemen mounted on quality horses of even pace in order for the accurate and synchronized movements that these drills entailed even to be attempted. These drills took months of practice to master, far more than a week of collective training allows for. This was, therefore, very challenging training for part time volunteers to undertake.

Parade drills would have been followed by mounted sword drill, under the tutelage of the Regimental Adjutant, always a regular permanent staff appointment. They would then march back to the town in the afternoon. By then the horses would have been actively ridden from 08.30 in the morning until after lunch and would have done enough for the day. The Regiment, having stabled the horses, then paraded in the market square for dismounted foot and sword drill.

Day 3: The Regiment would march out to the Training Ground for a repeat of the work of the previous day. That evening there would be an Officers' Ball in the Town Hall for their own and the Regiment's acquaintances and friends, the gallery decked, of course, with boughs of gorse in bloom, as well as other flowers. Supper served in the Council Chamber consisted of a sumptuous menu of which the following is an example:

Saumon à la Tartare

———

Poulet à la Broche, Poularde à la Duchess

———

Galantine de Veau

———

Hanche d'Agneau à la Anglaise

———

Pâté de Pigeon

———

Jambon et Lengue, Anchois à la Millionnaire

———

Mousse de Volaille en Aspic

———

Mayonnaise de Homard, Salad à la Francaise

———

Gelée au Madère, Crème à la Vanille
Charlotte Russe d'Apricot
La Bagatelle, Cha treuse d'Oranges
Meringues au Crème

The Regimental band produced the music for dancing afterwards. Sometimes the Ball was a joint Officers' and Sergeants' Ball. The Regiment of necessity consisted throughout its ranks of many relatively well-to-do people. The officers and other ranks often knew each other well, working and hunting together throughout the year.

Day 4: The Regiment paraded the following morning as usual, thick heads notwithstanding, and marched, with the band at its head, to the training ground for a training session under the eye of any friends or members of the public who wished to attend. The training started with a parade and inspection by the Commanding Officer followed by exercises in Troop Drills by each troop including attacks to the front, rallying to the right and left, retirement to the rear and rallies to the front once again. Dismounted fire action was then practised.

After an interval the Regiment then paraded under the Commanding Officer and practised further movements at regimental level followed by sword exercise and pursuing practices by word of command and in drill time, under the Adjutant. Finally the Regiment reformed and advanced in review order, a very difficult movement.

That evening there was a concert performed by the Band in the Market Square.

Day 5: The Regiment paraded as usual and marched out to practise outpost duties, by dividing itself in two, half to set up outposts, using the local river lines, canals or other prominent features, the other half of the Regiment to send out 'communicating patrols' and 'reconnoitring patrols' to test them. The features used by the Sherwood Rangers, for training, were used throughout the twentieth century, as well as the nineteenth century, to practise precisely the same tasks. The only difference was that the change in equipment brought a change in some of the technical terms, 'double vedettes' and 'cossack posts' being replaced by 'hides' and 'OPs'.

In the evening a smoking concert or 'smoker', as they are now known, was held. This was a convivial, informal evening of drinking, eating and entertainment, held in a local concert hall or large entertainment room in a local hotel hosted by the whole Regiment. Its purpose was to entertain the Regiment's families and friends and all within the local community who had made facilities available to the Regiment, or helped them in other ways during Camp.

The entertainment was of the home-grown variety, and performed under the genial control of one of the senior ranks, appointed Master of Ceremonies for the evening. This way of entertaining the Regiment's friends was introduced, it seems, over the years in preference to some form of Ball hosted by all ranks, which left the officers free to entertain more widely on another occasion. A typical programme included old favourites such as 'Soldiers of the Queen', 'Tommy Atkins', 'Charge of the Light Brigade' and 'Poor Terry of Derry', to name but a few, and ended with the National Anthem.

Day 6: (assuming that day to be Sunday) The Mayor and Corporation of the town, the Regiment, and often the local Volunteer Unit attended the Parish Church for divine service. Both units paraded in the Market Square and marched into church to the Regimental Band. The occasion always drew a large crowd as well as filling the church to overflowing. 'Onward Christian Soldiers' always featured and the sermon was seemingly always given by The

Reverend Sir Richard Fitzherbert, who held the appointment of Chaplain to the Regiment for many, if not all, the years of the second half of the century.

Day 7: One day was always devoted to a Military Tournament, keenly competed for by members of the Regiment, consisting of games and practices of military relevance. The location was the training ground and it always attracted a crowd of several thousand, who even paid to see the contest.

The sports included 'Lemon Cutting', slicing lemons on a post with the sabre at the gallop, 'Tent Pegging', lifting a tent peg driven into the ground on the point of a sabre whilst leaning down from the back of a galloping horse, 'Turk's Head, 2 Hurdles', which consisted of jumping two hurdles and slicing through the 'Turk's head', a swede or similar item with the sabre, 'Tug-of-War', which needs no explanation and the 'Victoria Cross Race', wherein the contestants race their horses down the course, dismount, pick up a simulated fallen comrade, remount, and race back up the course with him. There were other similar competitions in other years.

Throughout, the Band played music.

Day 8: Each year the Regiment, in common with all other yeomanry regiments, was inspected formally by the Inspector of Cavalry for the military district the regiment was in. In the case of the Sherwood Rangers this was the Northern District. His rank was full Colonel. The event took place on the Training Ground, again in the presence of a crowd of several thousand, including many distinguished members of the county and their wives.

The inspection commenced, once the Regiment was on parade and steady, by the arrival of the Inspecting Officer at the gallop accompanied by his orderly, reining in to a halt, to be received by a general salute. He then rode down the ranks, inspecting the Regiment closely for turnout of both horse and rider and their respective equipment and accoutrements. He always paid particular attention to the correct fitting of saddles, which, in the Regiment, was invariably of an exceptionally high standard.

The Regiment then, on command, initially marched past in single file, by sections and by Troops at the trot. This was then followed by a series of field manoeuvres consisting of the Regiment by Troops in advancing, wheeling, retiring and reforming at all paces up to and including the gallop. The Regiment then simulated throwing out a skirmish line on foot and engaging a simulated enemy with their carbines. Finally the Adjutant put the whole Regiment through a series of sword exercises. A report then followed.

After the Inspecting Officer had left, the Commanding Officer then presented the various prizes, the most prestigious of which was the Duke of Portland's prize for the best turned-out in the whole Regiment, and various Troop Prizes. The Regiment then marched back into the town and was dismissed. The final event was a promotions' conference conducted by the Commanding Officer.

The merriment and high spirits of the Yeomanry whilst at Camp used to manifest itself in a number pranks which were received with good-natured tolerance in the towns. The pranks took a number of turns. By way of example, one year in Worksop it was reported that at 3.00 on Sunday morning 'a large hat over the establishment of the Messrs Plants' was removed and placed on top of the pole of Mr Arthur Gilling's shop in Potter Street where it caused much amusement during Sunday (presumably during Church Parade). During Sunday night it was taken down and decorated with ribbons after the style of a recruiting sergeant's hat and was placed as an extinguisher on the lamp in front of the Royal Hotel. To keep up the fun, the crier was sent round offering a reward as to the authors of this outrage, the reward to be paid by one 'Jumbo' (an allusion to the presumed size of the hat's owner on account of the size of the hat itself). The pranksters' reply was a sign announcing a lecture from one Mr Verdant Green on the subject of 'Jumbo'.

Day 9: There was a final Review and the Regiment dispersed to the various Troop locations.

Accidents did happen to both men and horses. There was at least one member of the Regiment hospitalized by a fall each camp, broken bones being the most common, and occasionally horses were lost as well, due to a leg in a rabbit hole for example.

As can be imagined the impact of these Camps on the local community was significant, and a close relationship was developed between the Yeomanry and the towns' people who, as has been mentioned, turned out in thousands just to watch the Regiment march out to their Training Ground each day and back each afternoon, and attended the various parades and tournaments. It must have helped that it was calculated that a Camp in one of the towns would have been worth £5,000 in the monetary value of the time to the economy of the town, a considerable sum.

Chapter 26

Camps and Events

The following are some events which were highlights in the life of the Regiment in the last half of the nineteenth century.

In 1878 the Regiment had the honour, together with the South Notts Hussars, of escorting the Prince and Princess of Wales from Bestwood Lodge, the home of the Duke of St Albans, to Nottingham Castle, where they opened the new Museum. Happily enough, a detachment of 17th Lancers, who were at that time passing through the district, was also on that parade.

On assuming command in 1882 Lord Galway moved the Regimental Headquarters to Victoria Street, Retford because of its closer proximity to his home, Serlby. Previously it had been at Newark and Worksop.

In 1884 a Squadron took part in a field exercise with the South Notts Hussars in the Cuckney area.

In May 1886, when the Regiment was at Camp at Worksop, there took place the first of several inter-regimental training exercises between the Sherwood Rangers and 1st West Yorkshire Yeomanry, later to be called the Queen's Own Yorkshire Dragoons, who were at camp on Doncaster Racecourse. It is interesting that Lord Galway preferred to train with that regiment rather than with the South Notts Hussars. It may be that they camped at different times or not sufficiently close to each other, or it may just be that he knew the Yorkshire regiment better. At that time it was commanded by Earl Fitzwilliam.

These exercises took a similar form: the 1st West Yorkshire Yeomanry, at eight troops, twice the size of the Sherwood Rangers, moved south from Doncaster on a three-troop front to try and penetrate the Sherwood Rangers reconnaissance screen and take a specified location, in this case Sandbeck Park, Lord Scarbrough's home. Often, however, it was Serlby that was the objective. Not to put too fine a point on it, in the well-established ways of the yeomanry, the objective on each occasion was the location of the party that it was planned should take place afterwards.

These exercises were ambitious and valuable training in what has already been explained was the most important role of the yeomanry cavalry regiments, namely battlefield reconnaissance forward, and defensive screens, and the passage of accurate information in which skilled use of ground played a vital part. They were also high profile events attracting many spectators on horseback who, no doubt, used the day as an excuse to ride across country to improve their view. The exercises generated the minor triumphs and disasters between the units that all exercises do, which fuelled the banter in the messes and beer tents afterwards. One from this first exercise was of a section of the Retford Troop who got 'pounded' in a lane by a pincer movement from the enemy from either end. All seemed lost, but this, remember was the 'Tally Ho!' Troop, no less, and they, literally, at a single bound were free as they jumped the roadside fence over

A YEOMANRY ENCOUNTER AT SERLBY, IN NOTTINGHAMSHIRE.

1. Asking a Guardian of Law and Order the way to the scene of the battle.
2. Fair spectators of the fray.
3. The 'Sherwood Rangers' galloping to seize the entrance to Serlby Park.
4. Final capture of Serlby Park by the 'Yorkshire Dragoons'.

a place too big for any of their would-be captors to follow, and made good their escape over a particularly stiff line of country. No doubt the fences got higher, stiffer and more numerous in the telling.

The two Regiments trained together in 1890, 1892, 1893 and 1899.

In 1886 a detachment of the Regiment provided a Troop to keep the ground during an inspection of the South Notts Hussars by HRH the Duke of Cambridge, the Commander in Chief. They were complimented particularly on their smart appearance.

In 1888 there was an invitation ball held by the Newark Troop in the Town Hall.

In 1891 the Sherwood Rangers provided a travelling escort from Nottingham station to Wollaton Park; it was commanded by Lieutenant Bacon. The occasion was another inspection of the South Notts Hussars by HRH the Duke of Cambridge. The parade was also attended by Viscount Galway and Lieutenant T. R. Starkey.

In 1893 the Sherwood Rangers were themselves inspected by HRH the Duke of Cambridge, together with the Queen's Own Yorkshire Dragoons, now commanded by the 10th Earl of Scarbrough, whilst at camp at Worksop in May. The day of the inspection commenced with another in the series of exercises between the two regiments, the Yorkshire Dragoons, once again advancing south from their camp on Doncaster Racecourse, against a screen consisting of the Sherwood Rangers, 'resplendent in their green and gold uniforms and busbies'. This time the objective was Lord Galway's home, Serlby Hall. As usual, a large number of sightseers gathered from as far afield as Retford, Worksop and Doncaster. Not only that, but about 200 invited guests consisting of friends of the two regiments and representing most of what might be described as society of Nottinghamshire and south Yorkshire were present.

The newspapers reported that

HRH accompanied by a distinguished staff including General Sir Francis Grenfell KCB DAG for the Auxiliary Forces and General H C Wilkinson CB commanding North – Eastern District, rode out to witness the exercise and was rewarded by being a spectator of one of the prettiest bits of fighting off a battlefield.

The Dragoons, in due course, drew near to Serlby, the forward patrols of the Rangers having had them under observation for most of their advance. The right-hand squadron of the Dragoons

had advanced through Bawtry to the hill overlooking the gates of Serlby Park. Here the squadron had a warm reception from a dismounted troop of the

Rangers and would have suffered loss. The squadron however charged gallantly down the hill and in warfare had they jumped a somewhat serious obstacle (no match for the Tally Ho! Troop clearly) would have inflicted much damage on the skirmishers.

Meanwhile the main body of the Dragoons under command of the Earl of Scarbrough had moved across country on their left via Scrooby

with the evident intent of turning the flank of the Sherwood Rangers. A brilliant charge was delivered here and a certain number of Rangers were 'cut up' but a formidable obstacle in the Park gate intervening (Lord Galway himself no doubt) which Lord Galway was holding with a considerable force, the Dragoons had to withdraw with considerable loss.

The exercise, which was later praised by the Duke as having been done 'in a very creditable manner considering how little if any practice the regiments have hitherto had', concluded by midday and, after a break for luncheon, the two regiments were drawn up in review order in the park. It was a fine day, and

by then they bore little trace of the hard work they had undergone. By this time the attendance had largely increased, and the sight of the Yeomen in their glistening helmets, accoutrements, and gay uniforms was picturesque in

Insepection By HRH The Duke of Cambridge, 1893.

the extreme, while the galaxy of dress and fashion at the saluting base heightened the effect and produced an animated and brilliant spectacle.

HRH was greeted as he rode onto parade with a Royal Salute and

the regiments then trotted past by squadrons and ranked past by sections in excellent form and were subsequently put through a severe testing in field manoeuvres and their general bearing and efficiency was greatly admired.

The Duke later said

I have always had a strong opinion, which I retain more than ever, that the Yeomanry of this country are of great defensive power in the nation, but your special duties should they ever be required – which I hope this country shall never be in danger – but should an emergency or contingency arise the particular object for which you are formed in a military body is to avail yourselves of the experience and knowledge which your officers and men do have of their own localities in order to be the eyes of any army which may be

Garden Party at Serlby, 1893.

concentrated in the defence of the country. Therefore the drill I have seen this morning is to my mind the very thing, the very portion of the drill most adapted to the Yeomanry of this country.

The newspaper report then continued:

The party then adjourned to the lawn where a garden party was then held at which there was a large gathering of the local gentry. The regimental bands which had performed during the luncheon played at intervals and the proceedings throughout were highly pleasant and successful.

The centenary of the raising of the Sherwood Rangers was celebrated with great splendour during annual camp at Retford in 1894. This is an outline of the programme:

FRIDAY – A variety entertainment was given in the Town Hall by the officers of the Regiment. There were a company of dancing girls, under the direction of Miss Florence Bright, ventriloquial entertainments by Mr Charles King, musical sketches by Mr Geo Robbins, and the entertainment then merged at the close into an impromptu dance at which light refreshments were provided.

SUNDAY – The Regiment paraded in the Square at 10.30. A new standard was presented to the Retford Troop by the Viscountess Galway. The standard was

The Presentation of a new Standard to the Retford Troop by Viscountess Galway.
On the occasion of the Centenary of the Sherwood Rangers Yeomanry in 1894.

described as extremely beautiful, and as containing the initials of Viscount Galway (by whom the troop was formed), and of the Viscountess Galway, the Borough Coat of Arms, and other insignia, the whole beautifully worked. The Regiment then marched to the East Retford Parish Church, accompanied by the volunteers of both the Retford and Worksop companies. The preacher was the Chaplain of the Regiment, the Reverend Sir Richard Fitzherbert.

MONDAY –The Regimental Sports took place in the Park at Babworth. It was described as consisting of a 'most excellent programme'. In the evening the Retford Company of Volunteers, under the supervision of Lieutenant Danman, arranged a smoking concert in the Corn Exchange. It was under the musical direction of Mr John Smith.

WEDNESDAY – The Mayor (Mr Jno Hewitt) arranged to give a grand ball in the Town Hall to the Colonel of the Regiment, the officers, and others.

It is worth reproducing the guest list for this occasion because it provides an interesting snap shot of Nottinghamshire in the 1890s:

First the Officers of the Sherwood Rangers:

Colonel Viscount Galway, commandant; Major John Savile-Lumley; Captain and Adjutant J. S. Willett; Surgeon-Lieutenant C. Fleming; Captain Lieutenant Colonel Denison; Lieutenant R. C. Bacon; Major H. F. Huntsman; Lieutenant J. F. Laycock; Lieutenant H. Bromley; Lieutenant Peacock; Lieutenant M. S. Dawson; Captain F. H. Blacker; Lieutenant W. Hollins; Lieutenant T. R. Starkey.

Then the other guests:

Prince Adolphus of Teck
Viscountess Galway
General Keith Fraser
General Wilkinson, CB
Captain Owen
Captain Garrett
Lady Maud Rolleston
Miss Clayton East
Colonel and Mrs O'Shaughnessy
The Earl of Scarbrough, Sandbeck Park, and the officers of the Yorkshire
    Dragoons.
Sir Fredk Milner, MP and Lady Milner

Mr and Mrs Frank Huntsman, Forest Hill, Worksop, Mrs Graves Saule, and Miss M. Starkey

Colonel and Mrs Denison, Babworth Hall

Mr F. J. S. Foljambe (the Lord High Steward of the Borough), Lady Gertrude Foljambe, Mrs Arthur Duncombe, Miss Duncombe, Miss Bower, Miss Anderson, Captain Broderick and Mr Arthur Bower

Major Rolleston, Edwinstowe

Lady Ffolkes, Miss Ffolkes, Mr and Mrs Verelst, Mr Seymour, Mr Kidd, Miss Kidd, Miss Hannah Kidd, Mr E. A. Johnston and Mr W. H. Mason

Mr and Mrs Baines, Miss Baines, Miss B. M. T. Baines, Miss Nancy Baines, Mrs Crampton, Miss B. N. J., and Mr Ed. Baines, Mr Eustace Baines, Mr Woodley, Mr Ernest Whitehead, Mr Max Tylden-Wright, &c, the Hall, Bawtry

General Warrand, Westhorpe

Mr and the Hon. Mrs Jebb and party, Barnby Moor House

Mr J. F. Laycock, Wiseton Hall, Mr Fenwick, Mr Erskine, and Mr Clayton

Mr and Mrs Riddell and party, Hermeston Hall, Oldcotes

The Revd R. and Mrs Fitzherbert, Warsop Rectory

Mr and Mrs Robinson, Markham Hall, East Markham

Captain and Mrs Otter, Royston Manor, Yorkshire, Miss North, Miss Musters and Mr W. Fitzherbert

Mr, Mrs and Misses Bridgeman-Simpson, Babworth

Mr Gervase Beckett, Leeds

The Revd Watkin and Miss Homfray and party, West Retford Rectory

Lieutenant and Mrs Dawson, Ranby House, Mr J. A. Dawson

Mr A. Tylden-Wright, Mr H. Tylden-Wright, Miss Tylden-Wright, Miss Barnes, Mr Barnes and Mr Swannick

Mrs Seymour Randolph and Miss Garforth, Great Chesterford

Major and Mrs Willey, Blyth Hall, Mr and Mrs Dawson, Mr Gray, Mr Donisthorpe and Mr Mawson

Mr and Mrs John Robinson, Worksop Manor

Mr Sandford Robinson

Mr and Mrs Watkin Homfray, Beech Hurst, and Mr G. B. Hale

Mrs G. Butterfield and party, St Michael's House, Retford

Mr E. B. Faber, Harrogate and Mr W. Hood

Mr Thomas and Miss Hewitt, Weelsby Hall, Grimsby

Mr D. Elkington, London

Mr W. H. Wilkinson, Hawthornden, Gainsborough, Lincs

Mr J. D. Sandars, North Sandfield, Gainsborough

Mr A. E. Iveson, The Hollies, Gainsborough

Mr and Mrs W. Wright, Wollaton

Mr and Mrs Alderson and the Misses Hunter, Killamarsh, Yorks

Dr, Mrs and Miss Dawson, Saundby Grove
Mr, Mrs and Miss Lambert, Old Bank House, Retford
Mr and Mrs S. K. Marsland, Newark
Mrs and Miss Hart, Raven's Hill
Mr and Mrs Wilson, Reepham, Lincs
Mr Alderman Beeley, Retford
Miss Jessie Shilton
Mr and Mrs Joshua Walker, Babworth Road, Retford
Mr B. W. Stothert, Ordsall Rectory
Mr Hamilton White, North Terrace, Retford
Mr and Mrs Hopkinson, Thrumpton, Welham, Retford
Miss M. H. Pritchard, The Square, Retford
Mr S. and Miss Pegler, Amcott House, Retford
Mr C. J. E. Parker, The Bank, Grantham, Lincs
Mrs and Miss Hooker, Cannon Square, Retford
Mr and Mrs Dimock, Retford
Mr and Mrs Walker, Bawtry
Mr and Mrs Frank Pegler, Wood Leigh
Miss O' Gorman
Mr and Mrs Edward Wright, Ollerton
Mr Edwin Smith, Gringley-on-the-Hill
Mr D. Balfour, jun, Myre Hall
Mr Walter Smith, Gringley-on-the-Hill
Miss Nellie Park, Mr Tom Park, and Mr F. Park, Grove
Mr Walter Gibbs, Elkesley
Mr and Mrs Cottam, The Priory, South Leverton
Mr J. L. Bullivant, Torworth
Mr J. T. Simpson, Broughton, Lincs.
Mr and Mrs Foljambe, Brackenhurst
Mr and Mrs Samuel Roberts, Queen's Tower, Sheffield
Mr Hirst, Moorgate Villa, Retford, and Mr S. E. Hirst
Mr and Mrs P. H. C. Chrimes, Plumtree
Mr Alderman Curtis, Retford
Mr and Mrs Jesse Hind and party, Papplewick Grange
Mr and Mrs Oakden, Bank House, Retford
Mr, Mrs and the Misses Taylor Sharpe, Baumber Park
Mr and Mrs Lazenby, Park Villa, Retford
Mr and Mrs Calbert Appleby, Market Square, Retford
Dr and Mrs Tristan, Chapelgate, Retford
Captain Kelsey, Morton, Gainsborough
Mr and Miss Whitington, Tuxford
Mr and Mrs Wells, and Mr E. Wells, West Bank, Retford and Miss Ireland

Mr Percy Jones, Oaklands, Retford
Mr, Mrs and Miss Jones, Oaklands, Retford
Mr Leslie C. Cattle, Headon Manor
Miss Chapman, Elkesley Vicarage
Mr and Mrs S. W. Marshall and party, Lansdowne House
Mr J. W. B. Housley, Retford
Mr C. F. Elliott Smith, Mansfield
Mr A. B. Parke, Retford
Mr and Mrs R. H. Bate, Bridgegate House, Retford
Mr J. F. Jackson, Kersal Mount
Mr and Mrs W. A. Charles, Westfield, Retford
Mr and Mrs E. B. White, Arlington House, Retford
Mr and Mrs Kemp, Worksop
Miss Wilkinson, Clarkegrove Road, Sheffield
Miss Steel, Clarkegrove Road, Sheffield
Mr and Mrs Robert H. Allen, Holm Leigh
Miss Hind, Carolgate, Retford
Mr Herbert Marshall, Cleveland House, Gainsborough
Mr Hugh Marshall
Mr Chas A. R. Jowitt, Fulwood Road, Sheffield
Mr R. J. and Mrs Aynsley, Gosforth
Mr, Mrs and the Misses Skinner, Throapham Manor, Yorks
Mr Sam White jun., South Leverton
Mr H. Allcard, Sheffield
Mr J. Sharpe, Green Mile House, Retford
Mr Stanley
The Misses Wilkinson, Green Mile House, Retford
Miss Cattle, Welbeck House
Miss Peet, Brocco Bank, Sheffield
Mr and Mrs Bosvile, Ravensfield Park
Mr and Mrs Sorby, Endcliffe Crescent, Sheffield
Mr and Mrs Shaw, Carlton Hall
Miss Edith J. Haes, Bryntirion
Dr, Mrs and Miss Westbrook. Cannon Square, Retford
The Misses Hollingsworth
Mrs, Miss and Mr E. Conolly, Debdale Hall
Mr A. E. Rose, Middlecave House, Malton
Mr H. Rose
Mr and Miss Foottit, North Muskham Hall
Mr and Mrs W. Birks, Miss Nelly Birks, Miss Ethel Skinner, Mrs Stanley Birks,
    Mrs P. Birks and Mr W Thompson, The Hollies, Retford
Mr T. and Miss Eveline Macaskie

Dr and Mrs Thomson, Bridge House, Retford
Miss Lohden
Lieutenant Jno H Bradwell, Pelham Terrace, Nottingham
Mr F. L. Cutts, The Wharf, Retford
Mr and Mrs Goodbody, West Field, Retford
The Revd Geo. and Miss Shipton, Grove Rectory
Dr and Miss Thorne, The Square, Retford
Mr and Mrs H. W. O. Collingwood, Clayworth Hall
Mr Hodding, Worksop
Mr M. and Mr H. Van der Gucht, Worksop
Mr and Mrs Arthur Staniforth, Worksop
Mr R. H. Beaumont, County Club, Nottingham
Mr and Mrs Holmes, The Elms, Retford
Miss Twidale
Mr Motley
Colonel Newton, Hill Side, Newark
The Misses Knox, Bawtry
Mr W. Howett, Retford
Mr, Mrs and Miss Spencer, West Villa, Retford
Mr and Mrs George Peck, South Lawn, Retford
The Revd T. Gouch, Grammar School, Retford
Dr Trevor and Mrs Pritchard, Retford
Miss Munn, Miss Jane Munn, Gamston Rectory
Mr and Mrs J. G. Beevor, Barnby Moor; Miss Beevor, Miss Rose Beevor, Mr
    Harry Beevor, Mr J. I. Beevor, Miss Ellison
Mr Ingle Birkett
General and Mrs Keene, Miss Mary Ann Streatfield, Miss M. A. Scarlett, Miss
    Mabel Jebb
Mrs Fleming, Worksop
Mr and Mrs Geo. Walker, Dane's Hill
Mr, Mrs and the Misses Garforth
Mr J. H. Mawby, Retford
Mr A. Worthington
Mr and Mrs A. M. Eadon, Meadow Fields, Retford; and Mr and Mrs W. Eadon
Captain Penrose
Rev E. P. Sandwith
Mrs and Miss Huntsman, Harworth Vicarage
Miss Monson
Mr Hugh Huntsman
Miss Riddell
Mr Whitworth, Wath, Yorks
Mr Dawson, Thorncliffe,

Mr and Mrs Bernard Firth

Mr E. Dixon

Mr Gouldesborough

Lieutenant and Mrs Bromley

Miss Paget

Colonel Storey

Mr and Mrs H. J. Marchall, Gainsborough, Miss Marshall, Miss Ada Marshall

Mr R. A. Bradshaw, Mrs Bradshaw, Mr R. G. Bradshaw, Miss Bradshaw, Mr C.
W. Peace

Mr and Mrs Jacob Marshall, Grimsby

Mrs Farebrother, Grimsby

Mrs Sutcliffe, Grimsby

Mr Tom Sutcliffe, Grimsby

Captain Reed, Grimsby

Sergeant Park, Retford

Corporal Spencer, Retford

Mr and Mrs Henry Maxfield, Miss B. Johnson, Mr H. and Miss Denman

Mr George Kidd

Mr Chas H. Marshall

Mrs Caesar de Gusman

Miss Staniland

Mrs Fred Footitt

Mr John and Mrs McNaught Davis, Farndon

THURSDAY – The review took place which was followed by a Regimental
Dinner in the Town Hall during which Lord Galway spoke thus:

> After commenting on the position held by the yeomanry in the mobilizations
> scheme for the defence of the Empire his Lordship expressed the opinion
> that they were thoroughly able and ready to take their place when called
> upon. Let them hope that England would never lose her empire or her high
> position among the nations of the world and that their successors and in some
> instances their descendants would someday celebrate the bi-centenary of the
> Sherwood Rangers. He was sure of one thing – they could not be prouder of
> the green and gold than those who wore it at present nor would they be
> inspired by greater feelings of loyalty and patriotism.

The officers were quartered at West Retford Hall, the residence of Major
Huntsman, and each evening the Band played in the grounds during mess. They
also gave performances, under the direction of Bandmaster Holmes, in the
Market place, which attracted large companies of people and afforded much
delight to the residents.

On 20 May 1896 the Sherwood Rangers broke with tradition for the first time and camped in Nottingham in company with the South Notts Hussars. The reason was to train with their sister regiment for the first time, with whom they had been officially brigaded since 1893. The two regiments had been unable to work together before because there was no ground in Nottinghamshire large enough for two regiments to train on collectively. This had been overcome because Lord Middleton, the 9th Baron Middleton, had generously permitted Wollaton Park, already one of the South Notts Hussars' main training grounds, to be adapted for the purpose of this inspection, paid for privately by the South Notts Hussars.

The Sherwood Rangers mustered 173 out of 209 enrolled and the South Notts Hussars 335 out of 337. 'The officers of the two regiments had their Mess and Headquarters in the County House, High Pavement, the non-commissioned officers and men being mostly billeted in the town ...' Camp took the usual form, the key difference being that they did all their training jointly with the other regiment to work towards a joint inspection and review. There were joint sports and the two Regiments shared divine service in St Mary's church. The inspection was by General Luck CB, the Inspector General of Cavalry, for which the usual crowd gathered. It was said that it was a much larger crowd than for many years and this was attributed to 'the presence of the Sherwood Rangers, many of whose friends came from Grantham, Newark, and Mansfield, and even from Worksop and Retford districts'.

The review was most successful. Those who have witnessed it for many years were delighted with the smartness and general soldier like bearing of the men. There is probably no County able to boast so fine a brigade of cavalry as Nottinghamshire. The parade and field movements were carried out with admirable steadiness and on the whole there was this year an improvement in the mounts. The Sherwood Rangers who are a more privileged body with respect to horses than the South Nottinghamshire Hussars (many of whom reside in the county town) were especially admired; but there was certainly no fault to be found with the condition of the hussars. The general public owe a debt of thanks to the men who undertake hard training in order to be ready if needed for defensive purposes. On this account the remarks of General Luck will be read with no small degree of mortification.

He complimented both regiments on their turnout and appearance and the well groomed horses, but criticized them for the fitting of some of the saddlery and accoutrements and the over-ambitious nature of some of the manoeuvres attempted, given the inexperience of the two regiments at working as a brigade. Better to have truth than platitudes.

The last high profile event for the Sherwood Rangers of the nineteenth century was an inspection in 1899 whilst at Camp at Retford in May, as usual. The Inspecting Officer was the Commander in Chief, Field Marshal Viscount

Wolseley KCB GCMG, a famous fighting general, having enjoyed unbroken success in campaigns in Africa, against the Ashanti and Zulu, and in Egypt. He was so celebrated that he had been the model for Gilbert and Sullivan's 'Modern Major General'. It took a very similar form to that of the inspection by his predecessor HRH the Duke of Cambridge in 1893. Once again it was a joint inspection with the Yorkshire Dragoons at Serlby. It started with a similar advance by the Dragoons under the Lord Scarbrough, from Doncaster on Serlby, which was defended by the Rangers, under Lord Galway. This was followed by a review in front of a crowd of between 6,000 and 7,000 in front of the Hall. This is a remarkable number given that Serlby was well off the beaten track and most would have had to walk a mile or three from local railway stations. The following fulsome piece of journalism says it all:

Influenced by this rare climatic radiance and the joyous nature of the occasion, the ladies donned pretty costumes of cheerful colours many of exquisite taste – not only charming, but positively necessary in order to enliven the dull and sober habiliments of the men. The visitors completed three sides of a gigantic square the fourth being open to the woodland.

Within this expansive area the Brigade manoeuvred – the Yorkshire Dragoons in uniforms of dark blue with white facings, the Nottinghamshire Rangers in their well-known colours of green and gold, both regiments magnificently mounted upon horses of bay or brown or black – save a single grey, which looked oddly out of place – marching slowly or trotting freely according to order and in harmony with the changing music …, their swords and accoutrements dancing in the sunlight.

The bands of the two Regiments filled a dual role: they not only added to the beauty of the picture but they infused into the whole an element of cheerfulness, almost of hilarity, by their stirring military airs. And beyond, as a background to the fringe of people and to this centre of activity lay the sleeping woodlands, newly decked wholly or in part in the garments of Spring; above a glorious sky of deep azure blue adorned here and there with clouds of fleecy whiteness, floating almost motionless in the ethereal ocean: and below the velvety turf, soft and springy to the tread of a wonderful emerald green. It was a scene for the eye to revel in.

Its movements complete the Brigade wheeled into line and advanced in review order. When halted the Commander in Chief requested that the officers commanding regiments and squadrons should come to the front which they did. He addressed them with some remarks expressing pleasure with what he had seen

As the opinion of the Commander in Chief could not be heard as to what he had seen either by the crowd nor by the Regiments the writer may perhaps venture to give that of the general body of spectators, namely that in soldierly smartness and precision of movement, the Sherwood Rangers as compared with the Yorkshire Dragoons, bore away the honours.

After the parade there was food and water for the horses and lunch and beer for the men whilst Lord and Lady Galway entertained the officers and their guests in the Hall itself. The newspaper report is undoubtedly an exaggerated piece of work and thus almost gives the impression that the writer realized he was witnessing, as indeed, unknowingly, he was witnessing, the closing moments of a golden age. It was almost certainly the last time the Sherwood Rangers ever paraded as a full Regiment wearing their full-dress uniforms and their famous green and gold.

It is known the Commander in Chief was critical of regiments which had continued to be dressed as they had for the past century. In his view, it had become clear that khaki drill, introduced into the army ten years before, was the only appropriate uniform, having regard to the demands of modern warfare, but that was not the reason. The reason was the events unfolding in South Africa, of which the Commander in Chief alone amongst those present knew the detail, and which would, within weeks, result in the start of the Boer War. Not even in his wildest nightmares, however, could he have imagined that within eight short months of that glorious spring day a bunch of scruffy Boer farmers would have defeated and besieged one magnificently-equipped and trained and fully professional British Expeditionary Force, and tied in knots a second sent to rescue it.

Is also fair to say that, in the event of such an outcome, it is unlikely he had in mind that he would despatch the elegant bunch of amateurs now paraded before him, along with their fellow yeomen and many other volunteers, as reinforcements. It is fair to say that the elegant bunch of amateurs would have been no less surprised by such a possibility. Yet that is precisely what happened that Christmas. Possibly the one consolation, so far as the Commander in Chief was concerned, was that it would get them all dressed in khaki drill once and for all.

Now for a serious point with which to end this account of the first 105 years of the Sherwood Rangers Yeomanry. These Regiments had existed for a century and, due to an act of political judgement, for the last fifty years had been maintained as though frozen in time astride their beautiful horses, wearing the fine, but impractical, uniforms and performing the training movements of a form of warfare last practised fifty years before. They added up to a relatively large number of formed units consisting of civilian volunteers. The role amounted to little more than that of a general home defence reserve. In addition the equipment with which these units were provided was, at best, obsolescent and the majority was provided from the pockets of the officers and yeomen and not by the Government.

Yet no Government of whatever persuasion queried the policy and no one suggested a policy of justification based on a clearly identifiable need beyond that of an assumption that a standing reserve was valuable. The policy was simply that they were a cheap insurance policy against the unforeseen or the unforeseeable. Sadly the nineteenth century, a century of relative peace, changed at the precise moment just described, with a total lack of warning, to a new century of instability, uncertainty, fear and war, during which these units would distinguish themselves time and time again.

This creates a strong argument that this seemingly wasteful and illogical policy was in fact inspired. The question is why it worked so well? The answer could be that they were formed as complete units with their own leadership and infrastructure, identity and esprit de corps; they were self-motivated and were willing and able to respond to new threats or challenges and adapt to new roles and equipment quickly (as they did do in their thousands during Christmas 1899). This flexibility was born out of the fact that they were formed units with good basic unit skills. Surely the alternative approach to maintaining a standing reserve, which is to specify a role based on speculation concerning the future, is likely to fail since history teaches that it is impossible to second guess the precise nature of future needs. Trying to over-specify the role of a unit will result, more often than not, in that unit being ill-balanced for the task that in fact emerges which, according to the lessons of history, will be unforeseen and not those practised. In addition, the constant change and top-down micro management such a policy involves means that the unit is always of a smaller size than is needed to establish critical mass, and never has the chance to enjoy the kind of long-term stability which is key to producing units manned by volunteers with sound and effective levels of training across the broad band of basic military skills.

If no one could foresee either the Boer War or how it would develop a matter of weeks before it actually happened, how can anyone ever base a credible defence policy on an ability to foresee the future sufficiently well to tailor units to meet it? If, in addition, the role of the unit is primarily to produce individual reinforcements for the Regular Army, as is sometimes the case, the result tends to be that the raison d'être of the unit does not embrace everyone in it, but only those with the specific individual skills required. That is self-evidently divisive, bad for morale and, in the end, will tend to destroy the unit.

However, that latter description is of our current defence policy as it relates to reserves. Does it stand comparison with the policy as pursued in the nineteenth century? The conclusion that can be drawn is that the leaders of the nineteenth century pursued a policy, possibly by accident more than by judgement, which left the leaders of the twentieth century with a legacy which may well have been one of the differences between defeat and victory in the wars of their time. For what will the leaders of the twenty-first century be remembered in relation to their defence strategy?

As this history goes to print in 2011 a new policy, the first tailor-made for the Reserves in many years, has been tabled by the Government which, coincidentally, acknowledges many of the points made here. If this account establishes anything, however, it is that to convert rhetoric into action will not be easy, especially without adequate Government funding to replace the generosity of the Dukes and other private benefactors available in the past. But it is a welcome start.

Review of the Ducal Families during the Last half of the Nineteenth Century

Dukes of Norfolk

Having sold Worksop Manor to the 4th Duke of Newcastle in 1843, the Dukes of Norfolk had ceased to have any involvement in Nottinghamshire although, being based in Sheffield, the current Duke remained a close neighbour.

Dukes of Portland

The 4th Duke of Portland died in 1854 and was succeeded by his second son, William John Cavendish-Scott-Bentinck, born on 17 September 1800. The 5th Duke of Portland was an extraordinary man who lived a life of two distinct halves. In the first part he enjoyed a perfectly conventional involvement in public life and, in the latter, lived virtually as a recluse. He joined the Army in 1818 and spent the last ten years of his service in India before leaving as a Captain in 1834. Although he took the family seat in the Commons as MP for King's Lynn he had no interest in politics and surrendered his seat only two years later. Insofar as he supported any particular party he supported the Whigs. He was passionate about foxhunting: during the season he hunted literally every day, bar Sundays, maintaining a huge staff, a large stud of hunters at Welbeck and extensive kennels. He also maintained hunting lodges to which he moved his entourage and where he based himself to hunt country too far afield to reach from Welbeck in the day.

Once he succeeded to the Dukedom and to Welbeck, he made the development of the house and the estate his life's work. With a deep interest in agriculture he developed twenty-two acres of kitchen gardens, built a huge indoor riding school which contained a tan gallop over 400 yards long and many other houses and buildings. Most extraordinary of all, he built a number of underground tunnels linking separate parts of the estate, and even one leading to the railway station at Worksop. They were huge affairs, easily capable of taking a carriage and horses. Although preferring to avoid the company of outsiders he was regarded with great affection by people who worked on the estate who saw much more of him. By the time he died in 1879, however, the estate had fallen into disrepair. He never married and therefore left no heir.

If the 5th Duke shunned the limelight and a role in politics and public life, which had been such a part of the family's tradition, then his younger brother, Lord George Bentinck, more than made up for it. He was himself an MP and played a very prominent role, and was considered to be of great political influence nationally during the second quarter of the century. His career culminated in his promoting Benjamin Disraeli, one of the outstanding politicians of the century, on to the national stage and then to greatness. Their finest achievement together was, ironically, the defeat of Robert Peel over the hated Corn Laws, the very laws first introduced when Lord George's grandfather, the 3rd Duke, was in government, to protect the wealthy and aristocratic landowners' interests against the right to cheap food of popular demand.

Lord George's influence demonstrates that the Reform Act of 1832, which he, again ironically, supported changed much less in the early stages than its supporters hoped and claimed. What Lord George Bentinck achieved in sponsoring Disraeli to prominence was not dissimilar from the achievements of the Duke of Newcastle in relation to Walpole and Pelham and others over a century earlier, when few had a vote. This type of influence is still discernible even today. Equally, the true significance of the first Reform Act and the overriding influence of the popular vote is also now plain enough in terms of both its strengths and weaknesses.

Some say that the 5th Duke and Lord George worked together politically with Lord George as the front. That is open to debate. What is not open to debate is their great partnership in horse racing which was an enduring sport for them both.

The 6th Duke of Portland (William John Arthur Charles James) was the grandson of Lord William Charles Augustus Cavendish Bentinck, the third son of the 3rd Duke of Portland. He was born on 28 December 1857, acceded as 6th Duke of Portland in 1879 and died on 26 April 1943. Educated at Eton, he served in the Coldstream Guards, was granted many awards and held many appointments. Not only was he appointed KG but he was also the Chancellor of the Most Noble Order of the Garter, Lord Lieutenant of Caithness and of Nottinghamshire (1898–1939), Honorary Colonel of the 4th Battalion of the Sherwood Foresters, President of the Nottinghamshire Territorial Army Association, Provincial Grand Master of Freemasons, Nottinghamshire, and Bailiff Grand Cross of the Order of St John of Jerusalem, to name but some.

The 6th Duke was one of the most successful racehorse owners of his generation, winning the Derby twice and twenty-one other Group 1 races, fourteen Group 2 races, ten Group 3 races and two big Handicap races. He purchased the great stallion, St Simon, as a two-year-old, raced him successfully, and later stood the horse at Welbeck. It became the founding stallion of one of the most successful bloodlines in the world through its linear descendants Nearco, Neartic and Northern Dancer to the champion stallion Sadler's Wells and his sons Montjeu and Galileo. The Duke was Chairman of the first Royal Commission on Horse Breeding. He carried the Crown of HM Queen Elizabeth (the Queen Mother) at the coronation of George VI in 1937. He died in 1943 and was succeeded by his son William Arthur Henry Cavendish-Bentinck, Marquess of Titchfield (1893–1977) who commanded the Sherwood Rangers in the 1930s.

The Dukes of Newcastle

The 5th Duke of Newcastle was succeeded by his son Henry Pelham Alexander, born on 25 January 1834. He was, of course, as they all were, educated at Eton and then at Oxford, and sat as the Liberal MP for Newark between 1857 and 1859, which, considering that Newark had been a Conservative stronghold of the 4th Duke, would no doubt have set the latter spinning in his grave. He did not become directly involved with the Yeomanry, his interest being Freemasonry, but was nevertheless a great supporter. He was the Provincial Grand Master of Nottinghamshire Freemasons between 1865 and 1877. An obsessive gambler, a weakness which nearly destroyed the estate, he married on 11 February 1861 and his heir, Henry Pelham Archibald Douglas, was born on 28th September 1864 and succeeded him aged fourteen on his sudden death aged forty-five on 22 February 1879.

The 7th Duke was educated at Eton and Oxford and married Kathleen Florence Mary Candy, the daughter of a major in the 9th Lancers in 1889. Sadly, it was a childless marriage, but a very successful one nevertheless. He was a very small man in stature and had also lost a leg when he was young. Despite his physical disadvantages, he and his wife formed a strong team which did much to restore the fortunes of the Estate. Among many other interests, it was he and his wife who developed the Clumber spaniel breed of dog. He was a strong supporter of the Sherwood Rangers. At the coronations of Edward VII in 1902 and George V in 1911 he, by virtue of the lordship of the Manor of Worksop, provided a glove for the King's right hand and supported his right arm while holding the sceptre. This ancient right indicates the seniority nationally of the Manor of Worksop, derived from the ancient importance of Sherwood Forest as a Royal Forest.

The 7th Duke held various offices and directorships, the most prestigious of which was Knight of Grace of the Order of St John of Jerusalem. He was a Conservative, at last reverting to the family's traditional political stance and died aged sixty-four on 30 May 1928. Because he had no children of his own, his heir was his younger brother, Lord Francis Pelham-Clinton-Hope, who held a commission in the Sherwood Rangers between 1888 and 1892, and who succeeded him as the 8th Duke of Newcastle.

Earls Manvers

It was Sydney William Herbert Pierrepont who eventually succeeded as 3rd Earl Manvers on his father's death in 1860. He was the second son of the 2nd Earl Manvers, his elder brother having pre-deceased their father in 1850, having been born in 1825 when his father would have been forty-seven. Educated at Eton and Oxford, by 1852, when he would have been twenty-seven and the newly-titled Viscount Newark, consequent on the death of his elder brother, he was the captain commanding the Holme Pierrepont Troop of the South Notts Hussars. As has been seen, he was commissioned initially into the Sherwood Rangers and had transferred at a later date, probably on account of taking up residence at Holme Pierrepont. He was a very committed yeoman and commanded the South Notts Hussars between 1868 and 1879 and was appointed their Honorary Colonel in 1879. He became Conservative MP for South Notts from 1852 to 1860 when he succeeded to the title.

The 3rd Earl died in 1900 and was succeeded by his eldest son Charles William Sydney Pierrepont (1854–1926). He followed the family tradition by serving in the South Notts Hussars from which he retired as a captain but obviously continued to serve, probably in the Sherwood Foresters because he became the Honorary Colonel of the 8th Battalion, having been appointed Brigadier General Commanding the North Midlands Infantry. Elected as Conservative MP for the Newark division of Nottinghamshire in 1885, he served until 1895, and was re-elected unopposed in 1898. He was a hunting man and was Master of the Rufford for some years.

The Earls of Scarbrough and Baron Savile

It has been explained that on the death of the 8th Earl of Scarbrough the title reverted to Yorkshire due to the fact that his issue were all illegitimate, although his children continued to live at Rufford Abbey and, therefore, the following summary of the five illegitimate sons of the 8th Earl of Scarbrough may be of interest.

As can be seen the first George Frederick (1812–1816) died very young. The second John, who changed his surname from Lumley-Savile to Savile (1818–1896) never married but entered the Foreign Office in 1841 and had an exceptionally distinguished career, serving at a senior level in almost every major embassy during some very interesting times, and eventually as Ambassador Extraordinary and Plenipotentiary at Rome, for which he was ennobled as Baron Savile of Rufford with special remainder, failing heirs of his body, to John Savile-Lumley Esq., mentioned earlier as serving in the Sherwood Rangers. He was the only son of Frederick Savile-Lumley, Rector of

Bilsthorpe and was also highly regarded for his knowledge of fine arts, holding a number of prominent memberships of fine arts' academies, both at home and abroad, perhaps the most distinguished of which was his trusteeship of the National Gallery. He made a significant collection of antiquities during his life time, gifting some to the British Museum, and others to form the Savile Gallery in the Nottingham Castle Museum.

The third son was the Reverend Frederick Savile-Lumley (1819–1859) mentioned earlier. The fourth was Henry Savile-Lumley (later Savile) (1820–1881) to whom his father, the 8th Earl of Scarbrough, bequeathed the Savile family estates of Rufford and Thornhill, consisting of some 34,000 acres, of which 18,000 were in Nottinghamshire, provided he took the name Savile. He served in the Life Guards, retiring as a captain and made good use of his financial good fortune by becoming a very successful racehorse owner, his horses including Cremone which won the Derby, among other races, and the well-named The Ranger which was the first English-trained horse to win the Grand Prix de Paris; that must have warmed his French blood.

The fifth son was Augustus William Savile-Lumley (later Savile) (1828–1887), who is described as residing at Rufford. There is a strong indication that Rufford Abbey was the main base for Lord Savile when in England, because it is said that he greatly improved Rufford. He was a considerable collector of fine pictures and displayed them there.

It is worth noting the remarkable impact of these families on the affairs of the nation and the county. Quite apart from the many achievements in politics and public life generally mentioned already, and the fact that between them they owned most of the county, they were some of the foremost agriculturalists, foresters, and collectors of fine arts. They created the mining industry in Nottinghamshire and at least one great quarrying company at Steetley. They also had a major pastoral impact, playing a significant part in the beginnings of yacht racing as a sport, horseracing, hunting, shooting and golf. Their time may have been passing and their attempts to cling to power against the steady tide of Reform unhelpful, but their legacy is enjoyed by many still. Not only that but in most cases the descendents, adopting a much lower profile, still make a significant contribution to the county and to the Sherwood Rangers and South Notts Hussars.

Index

Note: Each Nottinghamshire based place name in the index has a grid reference which corresponds to the map of the county which appears in the endpapers.

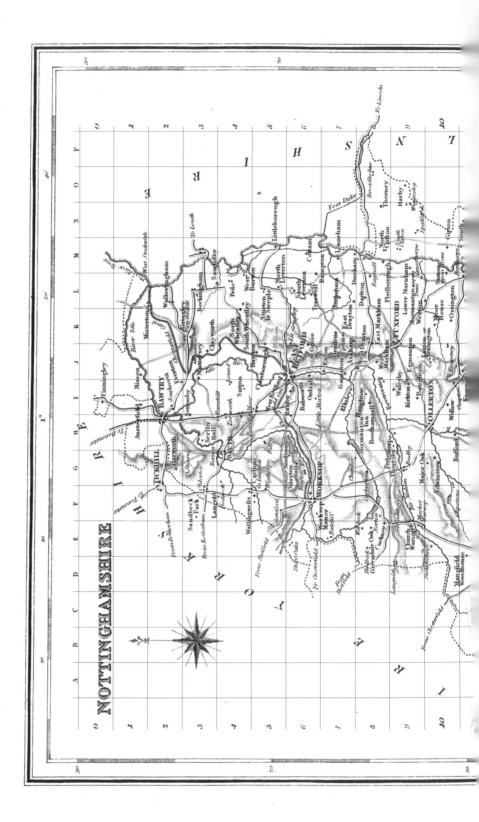

NOTTINGHAMSHIRE